**DIGITAL SIGNAL T
LINE CIRCUIT TEC**

DIGITAL SIGNAL TRANSMISSION

Line Circuit Technology

BRYAN HART

Formerly Senior Lecturer and Group Leader in Electronics
Department of Electrical and Electronic Engineering
North East London Polytechnic

VNR
UK

Van Nostrand Reinhold (UK) Co. Ltd

To my wife,
Rosalie Maria Hart-van Waard

First published in 1988 by
Van Nostrand Reinhold (UK) Co. Ltd.
Molly Millars Lane, Wokingham. Berkshire, England

Typeset in 10/12pt English Times by
Colset Private Limited, Singapore
Printed in Great Britain
at the University Printing House, Oxford

British Library Cataloguing in Publication Data

Hart, Bryan
    Digital signal transmission: line circuit technology.
    1. Digital communications
    I. Title
    621.38′043   TK5103.7

ISBN 0-278-00032-0

Contents

Preface

The coming of digital electronics has given rise to many textbooks – outstanding among which are the authoritative works of Douglas Lewin* – dealing, in the main, with the combinational and sequential logic aspects of system design.

By comparison, the coverage of digital hardware has been meagre: in particular, books on logic circuit interconnections have been scarce in number. When circuits operated at relatively low speeds this did not cause major problems to engineers and students, but the increasing use of fast, and very fast, logic circuits employed extensively in modern digital systems has brought the requirement for properly-designed transmission paths between printed circuit boards, on the boards, and even in the integrated circuit packages themselves.

This, in turn, has necessitated a physical understanding of line pulse behaviour, and an ability to design practically appropriate interconnection systems, by a far wider audience of engineers, scientists and students than has been the case hitherto.

For many years the study of transmission lines was, primarily, the province of the telecommunications engineer sending high-frequency radio signals to antennas for radiation into space, and the power engineer working at low frequencies (50/60 Hz) but with corresponding wavelengths comparable with the relatively large distances involved in the supply of power from generating stations to remote users. The textbook treatment of lines was well-established, but based mainly on the assumption of sinusoidal signals. This 'frequency-domain' approach is not best suited to the understanding of the transmission of digital signals.

This book is a primer covering the 'time-domain' approach to the propagation of pulse and digital signals along transmission lines. As it is an introductory treatment, attention is focused on the transmission of Transverse Electromagnetic Mode (TEM) signals along lossless lines.

---

*Published by Van Nostrand Reinhold (UK) Co. Ltd.

The book is intended as a text for use on undergraduate-level courses in electronics, digital systems, etc., at universities, polytechnics and colleges. The tutorial style of treatment – with worked solutions, to all end-of-chapter problems – has been adopted to make the book suitable, also, for self-study by those wishing to switch discipline, e.g. from physics to electronics, or to update their knowledge.

This volume is the result of many years spent in teaching a subject that students often find difficult at first encounter. It contains a number of approaches, some thought to contain a novel ingredient, intended to clarify the topics discussed.

The aim of Chapter 1 is to explain the difference between a lumped circuit and a distributed circuit in terms of a proposed time-domain response criterion.

Much of Chapter 2 is common property but special attention is paid to the sign conventions in, and the physical significance of, the equations derived. The concept of a 'Phantom generator', implicit in some literature, is made explicit here for use later in the book.

Chapter 3 considers the reflection chart and its applications, in depth, and includes the case of reflections on initially charged lines. The simple extension of the reflection chart to three dimensions and the use of the Phantom generator help to illuminate the progress and reflections of waveforms with non-zero rise and fall times, a topic often conveniently ignored.

Chapter 4 develops a physical background for the 'Sliding-Load-Line (SLL), or Bergeron, approach to the problem of reflections on lines with nonlinear terminations such as are met with in logic circuit interconnections. TTL and ECL are considered, in particular, but the technique is quite general in its applicability.

Time domain reflectometry waveforms (Chapter 5) are derived using the SLL and phantom generator concepts. Chapter 6 is a fundamental discussion of crosstalk and includes analogies intended to help in the understanding of forward and reverse travelling crosstalk waveform generation.

Chapter 7 is an introduction to some hardware aspects of 'bussing'. Its aim is to give the reader sufficient background information to be able to make intelligent use of the data sheets and application notes of the manufacturers of digital system components.

Appendix A is a résumé of practical line data. Appendix B gives details of practical circuit jigs, useful in experimental work, for those who do not have access to expensive equipment. Appendix C discusses practical problems met by engineers in the observation of waveforms with fast edges, e.g. ECL signals.

Most of the book can be understood without much mathematics. Such engineering mathematics as is included is there for completeness. The diagrams are meant to 'tell the story'.

Acknowledgements

Thanks are due to: Texas Instruments Ltd for permission to publish, in Chapter 7, abbreviated specifications for RS232C, RS485, abstracted from a technical brochure; Mr 'Bob' Pearson, friend and ex-colleague at NELP, for constructive critical comments on a first outline draft of the manuscript; Mr Paul Costa for translating my rough sketches into neat line drawings; Mr Mark Corbett and his colleagues at Van Nostrand Reinhold (UK) for acting as 'midwives' for the book; my wife, Rosalie, for her patience and understanding when a periodic inability to choose the 'right' word caused me to be less than agreeable.

Notation

Unless otherwise specified in the text, the symbol convention used in this book is as follows.

DC, *bias* and *peak amplitude* quantities are represented by an upper-case symbol, with an upper-case subscript if required. Thus, V_{CC} is the electro-motive force (e.m.f.) of a bias source, V the constant potential difference (p.d.) across, say, a resistor.

Small signal quantities representing changes are shown by a lower-case symbol with a lower-case subscript: thus, v_f is the incremental forward travelling wave on a line.

Total instantaneous quantities are represented by a lower-case symbol with an upper-case subscript, e.g., v_G for pulse-generator output.

The opening and closing of a switch, connected to a line, is assumed to occur at a reference time $t = 0$: the notation $t = 0 +$ is used to refer to conditions an infinitesimal time after the switching action has occurred.

1 Lumped and distributed circuits

In conventional circuit theory, which deals with the electrical performance of ideal circuit elements (e.g. an ideal resistor, etc.) and circuits formed by interconnecting them, 'cause' and 'effect' are regarded as time-coincident. This fundamental concept is extended to include the world of the real component, i.e. the physical hardware whose electrical performance over its useful range of operation approximates closely to that of the ideal element. Thus, the application of a voltage source to a resistor is assumed to affect, *simultaneously*, all points within the component. The mechanical dimensions of the resistor affect only the magnitude of a time-independent proportionality factor R, the resistance, relating the potential difference v and the current i: the component is considered 'lumped'.

It is normally valid to make this assumption of simultaneous action at all points because the propagation of electrical disturbances is known to occur with a velocity approaching that of light ($\sim 3 \times 10^8$ m/s, in air). Consequently, any electrical disturbance, however rapid, at one end of a resistor 1 cm long makes itself felt not later than about 0.03 ns at the other end. This delay may be regarded as negligible for pulse signals having rise and fall times exceeding a few nanoseconds.

For many branches of electronics, lumped circuit theory is appropriate in analysis and design. There is no need to resort directly to the classic equations of electromagnetic theory formulated by James Clerk Maxwell. However, a circuit must be regarded as 'distributed', rather than lumped, if a delay between cause and effect becomes significant, i.e. the equations describing electrical behaviour involve both space and time variables.

A practical criterion by which a system can be judged as lumped or distributed for the purposes of digital electronics is now discussed.

In Fig. 1.1 a generator producing a signal known as a truncated-ramp function voltage waveform is connected by two closely spaced parallel wires to a remotely located load resistor R, the mechanical dimensions of which

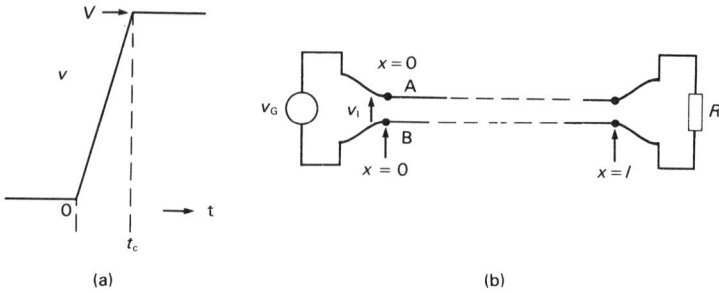

Fig. 1.1 Circuit for defining a distributed system. (a) Generator wave-form, a truncated-ramp function; (b) Line and remotely located load resistor R.

will be neglected in this discussion. The two wires, whose length, l, greatly exceeds their spacing, comprise an elementary 'transmission line' — so-called because of its role in transmitting the signal from source to load.

Now the time taken for an electromagnetic disturbance to travel to the right from the terminals AB to the load R is defined as the 'one-way transmission time' and will be designated by the symbol t_d. It follows that $t_d = (l/u)$, if u is the velocity of signal propagation ($\approx 3 \times 10^8$ m/s); alter-natively, $t_d = l t_u$ in which $t_u(= 1/u)$ is defined as the time-delay per-unit-length. (t_u is assumed to be a constant, independent of l, for a given line. See Chapter 2.)

If $2t_d \ll t_c$ the two-way transmission time is much less than the signal transi-tion time. No significant signal delay is involved. The signal source is influenced by R while the signal is changing. This is the case of a 'short-line' and traditional lumped circuit theory can be applied.

Suppose, instead, that $2t_d \gg t_c$. The circuit must now be regarded as distributed because signal delay cannot be ignored. The input voltage generator cannot be affected by R until a time $2t_d$ has elapsed. This is the case of a 'long line'.

The limit condition $2t_d = t_c$ is an arbitrary, but convenient, criterion for judging whether a circuit or system is to be taken as lumped or distributed.

Fig. 1.2 shows pictorially that, in the classification of systems, it is not so much the time required for signal transmission that is important as its magnitude in relation to the fastest transition times in the signal waveforms transmitted. In Fig. 1.2 the vertical and horizontal axes have the same scale (in, for example, ns/cm) for the two variables involved. The line OC, with a slope $+1$, describes the limit condition $2t_d = t_c$ and serves to distinguish between lumped and distributed systems.

In engineering calculations the condition $a \gg b$ is usually safely met if a and b differ by an order of magnitude (though a factor of 5 is often accept-able in practice), so for $a \gg b$ we write $a \geqslant 10b$. Accordingly, the line OL with a slope $+10$ serves to define a region of the plot, shown horizontally shaded,

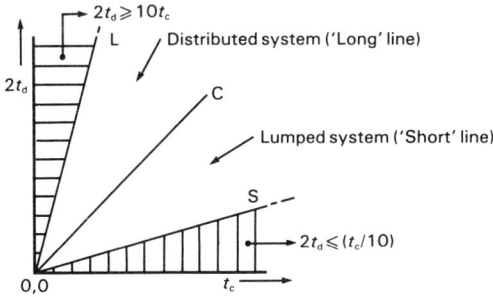

Fig. 1.2 Pictorial classification of lumped and distributed systems.

for which $2t_d \gg t_c$. In this region the 'transmission line' must have uniform and reproducible characteristics if predictable signal voltage waveforms are to be developed across R. The line OS with a slope $+0.1$ serves to define a region of the plot, shown vertically shaded, for which $2t_d \ll (t_c/10)$. In this region the nature of the wiring between components and other equipment is relatively unimportant and open 'hook-up' wire is often acceptable.

We have chosen the truncated-ramp voltage waveform of Fig. 1.1(a) as our test signal for classifying a system as lumped or distributed. This may seem arbitrary. Actually, it is a good engineering choice. A perfect step-function waveform (corresponding to $t_c = 0$) does not exist in practice, though it is used in mathematical analysis. However, some practical wave-form edges do have small values of t_c and with a truncated-ramp waveform we can allow for whatever t_c is appropriate.

There is another good practical reason for choosing a waveform such as that in Fig. 1.1(a): the edges of most logic circuit waveforms approximate truncated-ramps over their useful regions of operation, viz. those sections responsible for switching logic gates.

In concluding this introductory chapter we have to decide how to classify the system when a waveform such as that of Fig. 1.3 is applied to a transmission line. The waveform of Fig. 1.3 can occur in some forms of time-base sweep circuit. The aim is to generate a linear voltage sweep characterized by $v = Kt$, where $K(>0)$ is time-independent, and $t_s \geqslant t > 0+$, following a switching action at $t = 0$. This is part 's' of the waveform.

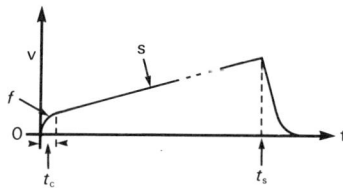

Fig. 1.3 Waveform for which both lumped and distributed analytical circuit techniques may be required.

However, owing to circuit design limitations, there is a small jump in voltage — part 'f' of the waveform — occupying a short time interval t_c. Suppose $t_c = 10$ ns, $t_s = 5$ μs and $t_d = 20$ ns. For these conditions we can regard the system as lumped for part 's' of the waveform and apply the standard techniques of lumped-circuit theory. But, we must regard the system as distributed for section 'f' of the waveform and apply the theory and practice expounded in the remainder of this book.

Finally, we have used a parallel-wire transmission line in the discussion so far, and will continue to do so except where otherwise indicated, for ease in drawing. However, the argument is equally valid for other standard forms that are more convenient in practice. These are the twisted-pair, coaxial cable and microstrip lines. (See Appendix A for their symbolic representation and a résumé of their characteristics.)

PROBLEMS

P1.1   For the system of Fig. 1.1, $l = 3$ m, $u \approx 3 \times 10^8$ m/s.
Estimate a value for t_c for which:
a conventional circuit theory is applicable;
b conventional circuit theory is definitely not applicable.

P1.2   A system such as that of Fig. 1.1 has $t_c = 1.5$ ns, and $t_u = 5$ ns/m.
Show that $t_c = (2l/u) = (2lt_u)$, if $l \approx 6$ in.

P1.3   Construct a classification diagram similar to Fig. 1.2 but with the vertical axis scaled in metres. (Assume $t_u = 5$ ns/m.)

2 Characteristics of ideal transmission lines/cables

To avoid unnecessary repetition, the words 'line' and 'cable' will be regarded as equivalent for electrical purposes, though their physical constructions may vary widely in practice.

We will define an *ideal line* as one which merely delays a signal by an amount that is calculable and directly dependent on its length. Though the velocity of propagation (u) and line length (l) are sufficient to define the one-way delay time $t_d = (l/u)$ it is often more convenient to specify the time-delay per-unit-length $t_u = (1/u)$.

There is no distortion of waveshape or change of amplitude for a signal travelling on an ideal line. As a signal progresses along it, in a given direction, there is a time-independent factor relating line voltage and line current. This is the characteristic impedance Z_0 which, for an ideal line, is a real-number parameter with the dimensions of resistance. The electrical performance of an ideal line is summarized in Fig. 2.1.

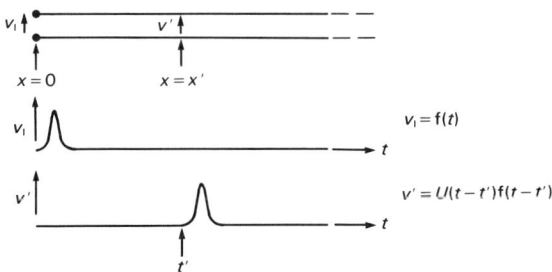

Fig. 2.1 Electrical characterization of an ideal line.
$$U(t-t') = 1 \quad \text{for } t > t' \,(= x'/u)$$
$$U(t-t') = 0 \quad \text{for } t < t'$$

Mathematically, $U(t)$ represents the Heaviside unit step function having the properties $U(t) = 1$, $t > 0$ and $U(t) = 0$ for $t < 0$. $U(t - t')$ is thus the *delayed* step-function having a value unity for $t > t'$ and zero for $t < t'$.

Hence, $v' = U(t - t')f(t - t')$ is merely a convenient mathematical description of a replica of the input signal $v_1 = f(t)$, delayed by a time interval t'.

An additional feature of an ideal line is that a pulse on it does not affect, and is not affected by, signals on nearby conductors.

2.1 EQUATIONS FOR AN IDEAL LINE

Consider the two-wire open transmission line of Fig. 2.2. It comprises two parallel wires closely spaced in air and is designated 'semi-infinite' as it extends from $x = 0$ to $x = \infty$. The electric (E) and magnetic (H) fields for a short section of line located between two perpendicular planes at x and $(x + \delta x)$, are shown in crude schematic form only; δx can always be chosen small enough for this section of line to be modelled by an appropriate arrangement of single lumped components. Two choices are shown in Fig. 2.3(a) and Fig. 2.3(b), for a general line. The parameters that characterize the line are L, R, C, G, which are respectively the per-unit-length values of inductance, loss resistance, shunt capacitance and leakage conductance. We focus attention, henceforth, on the case of a 'lossless' or 'loss-free' line, having $R = G = 0$, because this is often closely approximated in pulse work and the analysis of line operation is not obscured by second-order effects. Furthermore, we choose to analyse Fig. 2.3(c) rather than an equivalent based on Fig. 2.3(b) because Fig. 2.3(b) can give the misleading impression that one wire of the two-wire line is necessarily different from the other. Thus in Fig. 2.3(c) the line inductance is shown as distributed equally between the two conductors.

The special case of a line with finite R and G but no distortion is covered later in a problem (P2.8). For the small section of lossless line modelled in

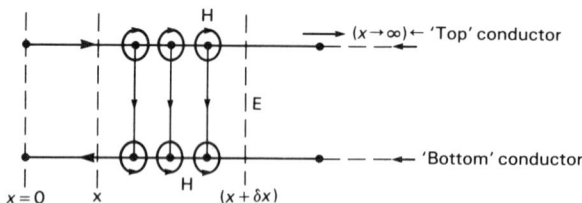

Fig. 2.2 Semi-infinite open parallel two-wire transmission line with associated electric (E) and magnetic (H) fields between planes at x and $(x + \delta x)$

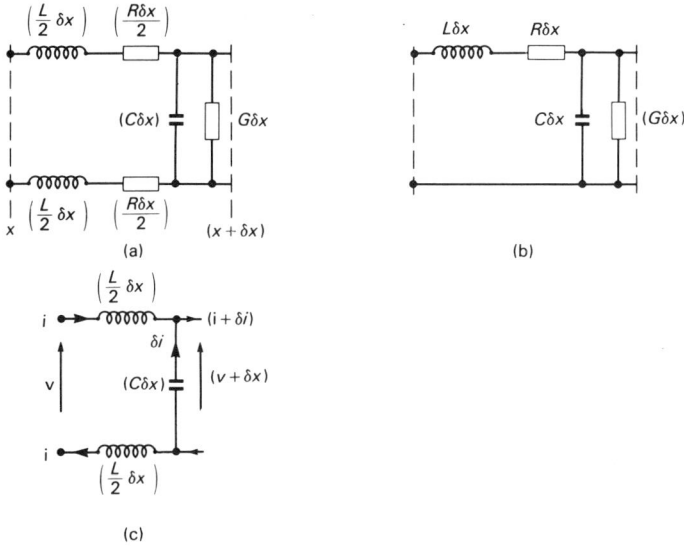

Fig. 2.3 Lumped model representation of line section in Fig. 2.2. (a) General line with losses; (b) Alternative representation of (a); (c) Representation of an ideal (lossless) line section.

Fig. 2.3(c), δv exists because of an inductive voltage drop between its ends and δi exists because of capacitive current.

$$\therefore v - (v + \delta v) = \{(L/2)\delta x + (L/2)\delta x\}(\delta i/\delta t) \tag{2.1}$$

$$\text{or, } -\delta v = L(\delta x)(\delta i)/\delta t \tag{2.2}$$

$$\therefore -(\delta v/\delta x) = L(\delta i/\delta t) \tag{2.3}$$

In the limit case this gives,

$$-(\partial v/\partial x) = L(\partial i/\partial t) \tag{2.4}$$

Partial differential coefficients are necessary here as v is a function of two variables, x and t.
Similarly,

$$i - (i + \delta i) \approx C(\delta x)(\delta v/\delta t) \tag{2.5}$$

which, in the limit, yields

$$-(\partial i/\partial x) = C(\partial v/\partial t) \tag{2.6}$$

From differential calculus,

$$\partial/\partial x\{(\partial i/\partial t)\} = \partial/\partial t\{(\partial i/\partial x)\} \tag{2.7}$$

The same relationship holds for v if it replaces i in eqn (2.7). Partial differentiation of eqn (2.4) with respect to x and eqn (2.6) with respect to t, and the use of eqn (2.7) gives,

$$(1/LC)(\partial^2 v/\partial x^2) = (\partial^2 v/\partial t^2) \tag{2.8}$$

By a similar procedure,

$$(1/LC)(\partial^2 i/\partial x^2) = (\partial^2 i/\partial t^2) \tag{2.9}$$

These partial differential equations define the line performance. They are linear because the product LC is independent of v, i, t, so any two possible solutions can be combined to give an equally valid solution.

One solution to eqn 2.8 is,

$$v = f\{t - x\sqrt{(LC)}\} \tag{2.10}$$

f can be any function, e.g. $v = k\{t - x\sqrt{(LC)}\}^2$ where k is a constant, but it is most likely that f will have the form of a truncated-ramp function voltage waveform or a decaying exponential voltage waveform.

Equation 2.10 describes an undistorted waveform of arbitrary shape travelling in the direction of increasing x. It is equivalent in form to,

$$v = U(t - t')f(t - t') \tag{2.11}$$

where $t' = (x'/u)$, and u is a constant propagation velocity given by

$$u = 1/\sqrt{(LC)} \tag{2.12(a)}$$

t_u the time delay per unit length is $(1/u)$,

$$t_u = \sqrt{(LC)} \tag{2.12(b)}$$

An interpretation of eqns (2.10) and (2.11) is given in Fig. 2.4. An alternative viewpoint in explaining eqn (2.10) is to imagine an observer travelling on the waveform so that, for him, the amplitude of v is constant, i.e. $f\{t - x\sqrt{(LC)}\}$ is constant. This implies that $\{t - x\sqrt{(LC)}\}$ is also constant. Differentiation then gives $dx/dt = u = 1/\sqrt{(LC)}$.

An important feature in waveform progress is the incremental impedance

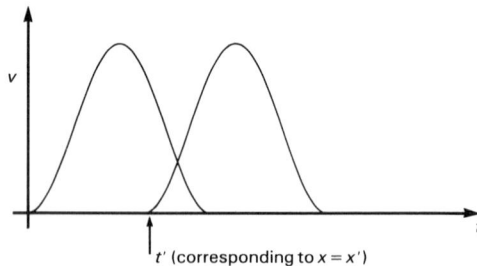

t' (corresponding to $x = x'$)

Fig. 2.4 Graphical interpretation of eqns (2.10), (2.11)

'seen' looking to the right between the wires at any point $x = x'$. This is the 'surge impedance'.

Dividing each side of eqn (2.4) by the corresponding side of eqn (2.6) gives,

$$(\partial v/\partial i)^2 = L/C$$
$$\therefore \ (\partial v/\partial i) = \sqrt{(L/C)} = Z_0 \qquad (2.13)$$

Z_0 = 'Characteristic impedance'. For the lossless line considered this is purely resistive and often designated R_0 and called the 'characteristic resistance.' R_0 and Z_0 will be used interchangeably for a lossless line.

R_0 does not, of course, represent a resistance that is dissipating energy by converting it to heat; rather it represents the ratio (v/i) involved in the transfer or transmission of energy *along* the line. A similar relationship to that of eqn (2.13) occurs in the shock-excitation of the initially uncharged loss-free parallel LC circuit in Fig. 2.5.

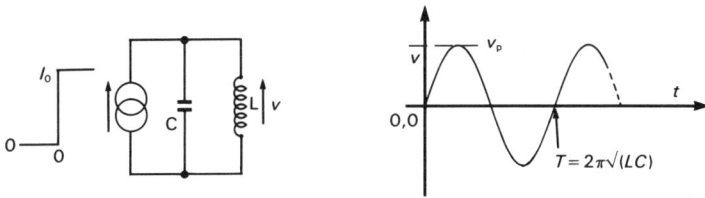

Fig. 2.5 (a) Shock-excitation of a loss-free parallel LC circuit. (b) If maximum value of $v = v_p$, then $(v_p/I_0) = \sqrt{(L/C)}$

After a step-function of current, I_0, is applied at $t = 0$ there is a constant exchange of energy between L and C. When, for any $t > 0+$, $v = 0$ all the energy $(LI_0^2/2)$ is stored in L; when the current in L is zero, v reaches its maximum value v_p and the energy stored in C is $(Cv_p^2/2)$.

As there is no energy loss,

$$(Cv_p^2/2) = (LI_0^2/2) \qquad (2.14)$$

$$\therefore \ (v_p/I_0) = \sqrt{L/C} \qquad (2.14(a))$$

(v_p/I_0) has the dimensions of resistance but no energy is dissipated. The quantity $\sqrt{(L/C)}$ governs the relationship between a potential difference and a current in an energy transfer process.

Note that the partial derivative in eqn (2.13) can be replaced by a total derivative because the relationship is independent of x, t. This is always true with a uniform line in which any two equal-length sections, selected at random, are indistinguishable in all respects. For instantaneous *total* quantities we can integrate eqn (2.13) to give,

$$v = iR_0 + \text{constant} \qquad (2.15)$$

The constant term refers to initial line conditions: it can, and will, be taken as

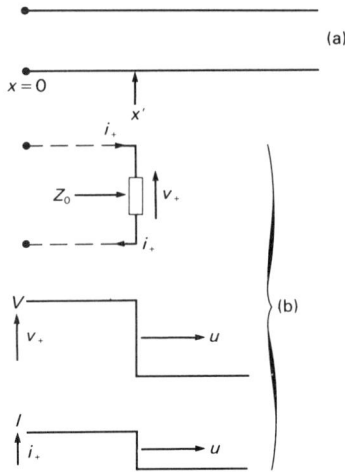

Fig. 2.6 (a) Initially uncharged two-wire line has step applied at $t = 0$; (b) Line conditions at $t' = x'/u$.

zero for representing *changes*, i.e. when v and i are accompanied by a subscript that is a lower-case letter or a plus or minus sign.

Considering, then, changes and using the subscript ' + ' to denote a waveform travelling in the direction of increasing x in Fig. 2.6, the incremental resistance at a point x' on the line seen looking to the right is R_0 where,

$$(v_+/i_+) = R_0 = \sqrt{(L/C)} \tag{2.16}$$

A point to be stressed is that v and i for the line as a whole are functions of two variables x, t. It is important to distinguish between sketches of distributions of v, i in space and variations of v, i with t. It is convenient to take either a moment 'frozen-in-time' and display conditions all along the line or, for a fixed point on the line, display the dependence of v, i, on t.

Figure 2.6(b) shows the line conditions for all x for an input step of magnitude V applied at $t = 0$ to an initially uncharged line. As the step progresses at constant velocity, u, v_+ and i_+ are 'in-step' as are their associated electric (E) and magnetic (H) fields. Actually, the conditions existing in Fig. 2.6 can be used to derive, somewhat less rigorously than previously, the basic relationships for t_u and (v_+/i_+) given in eqns (2.12(b)) and (2.16), respectively.

Thus, when the step waveforms have reached point x' the charge supplied by the input source V is $i_+(t_u x')$ and this must be equal to the charge $(Cx'v_+)$ accumulated by the line.

Hence, $i_+ t_u = Cv_+$

Furthermore, v_+ at x' can be regarded as generated by the rate of change of magnetic flux.

Hence, by the law of induction,

$$v_+ = (Lx')i_+/(t_u x') = Li_+/t_u$$

From these two relationships,

$$(v_+/i_+) = \sqrt{(L/C)} \text{ and } t_u = \sqrt{(LC)}$$

It can be shown that the energy of the wave is equally shared between the E and H fields (Problem 2.5).

Difficulties rarely arise when considering the progress of step waveforms on lines, but care has to be taken when considering waveforms of other shapes. Thus, consider the truncated-ramp waveform in Fig. 2.7(a). By the time that the input voltage has risen from 0 to V, in a time interval t_c, the waveform has started to travel down the line. Fig. 2.7(b) and (c) show the conditions on the whole line for two conveniently chosen times, respectively $t = t_c$ and $t = 2t_c$. This may be shown by regarding the ramp as the summation of a series of minute equal-amplitude voltage steps each delayed by the same time increment from its predecessor.

So far we have discussed only a forward-moving waveform corresponding to one solution of eqn (2.8). However, there are, mathematically, equally valid solutions of the form,

$$v_- = g\{t + x\sqrt{(LC)}\} \tag{2.17(a)}$$

or more generally,

$$v_- = g\{t + x\sqrt{(LC)} + X\sqrt{(LC)}\} \tag{2.17(b)}$$

where X is a constant that can be positive, negative, or zero. By an argument similar to that for the wavefront travelling in the direction of increasing x, these correspond to waveforms travelling in the direction of *decreasing x* along the line, with velocity $u = 1/\sqrt{(LC)}$, hence the subscript ' – '.

Consider a step voltage waveform (Fig. 2.8). i_- indicates the physical direction of current flow and, as for the forward travelling waveform,

$$(v_-/i_-) = R_0 \tag{2.18}$$

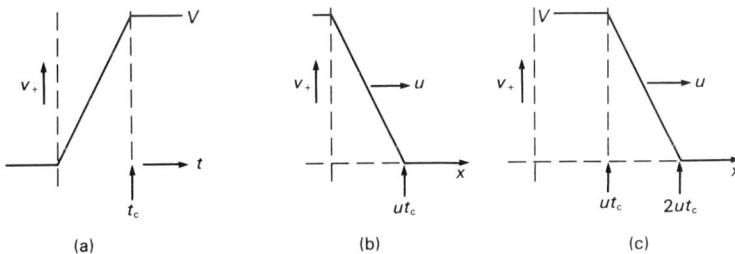

(a)   (b)   (c)

Fig. 2.7 Truncated-ramp voltage waveform. (a) $v_+ = f(t)$ at $x = 0$; (b) $v_+ = f(x)$ at $t = t_c$; (c) $v_+ = f(x)$ at $t = 2t_c$

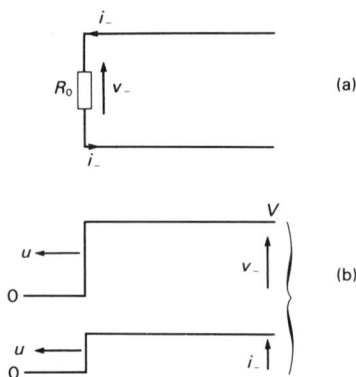

Fig. 2.8 A wavefront moving in the direction x decreasing. (a) The impedance seen between the wires for the wavefront is R_0; (b) Line conditions for a step wavefront of magnitude V on an initially uncharged line

For any further quantitative treatment of line pulse description we must adhere to a consistent sign convention and a long-established one is this. A positive line voltage is one for which the 'top' conductor in Fig. 2.2 is positive with respect to the 'bottom' one, and a positive current is one flowing from left to right in the top conductor (and hence right to left in the bottom conductor).

On this basis v_+, v_- are both positive. Let us write $v_f = v_+$ and $v_r = v_-$, where the subscripts f, r denote the forward and reverse waves respectively.

Then,

$$(v_f/i_f) = R_0 \tag{2.19}$$

In describing the reverse wave by a subscript r we have $v_r = v_-$ but, now, $i_r = -i_-$.

Thus eqn (2.18) becomes,

$$(v_r/i_r) = -R_0 \tag{2.20}$$

It must be emphasized that the minus sign in eqn (2.20) arises solely from the mathematical consistency of taking as positive, in sign, currents flowing from left to right in the top wire; the reader must not be misled by thinking that a negative resistance appears when a wavefront travels in the reverse direction!

It has not been established, yet, how or why a reverse waveform could occur when a waveform generator is connected to the line at $x = 0$. The mechanism that permits this possibility is dealt with in the next section.

For the present we can write as general solutions for eqn (2.8), (2.9), respectively,

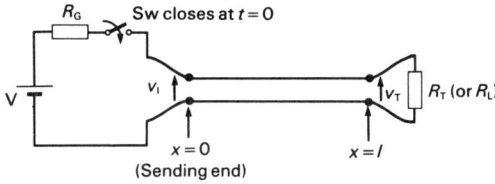

Fig. 2.9 A transmission line with voltage step-function drive

$$v = (v_f + v_r) = (v_+ + v_-) \tag{2.21}$$

$$i = (i_f + i_r) = (i_+ - i_-) \tag{2.22}$$

We note that when v_f and v_r both exist at a given point on the line, $(v/i) \neq R_0$.

2.2 POSSIBILITY OF LINE PULSE REFLECTIONS

Consider Fig. 2.9. A uniform, uncharged, lossless transmission line, of length l, having a characteristic resistance R_0 and 'terminated' by a load resistor R_T is connected at $t = 0$ by a perfect switch, Sw, to a source V in series with a resistor R_G.

Looking in at its input terminals at $t = 0+$, i.e. an infinitesimal moment after Sw has closed, the uncharged line appears as a resistance R_0 irrespective of the magnitude of R_T. A forward wave starts to travel down the line and the instantaneous line voltage at the wavefront is equal to the amplitude of the forward wave at $x = 0$. No reverse wave is yet possible: the forward wave must first 'discover' what exists at the end of the line. In other words, it must reach the end of the line before it can be affected by what is there.

Figure 2.10 shows a simple equivalent circuit for calculating $v_f = v_I(0+)$.

$$v_f = R_0 V/(R_0 + R_G) \tag{2.23}$$

· When the forward waveform reaches the end of the line there is no reflected

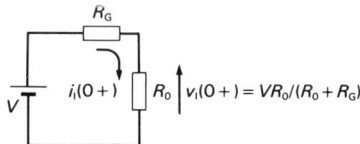

Fig. 2.10 Equivalent circuit at line input at $t = 0+$

waveform if $v_T(t_d) = v_f$. Thus the energy supplied by V in a time increment δt, following t_d, is $(v_f i_f)\delta t$. The energy dissipated in R_T in the same time interval is $(v_T^2/R_T)\delta t$. As $i_f = v_f/R_0$ it follows that if $R_T = R_0$ then $v_T = v_f$, all the incident energy is absorbed by R_T, and no reverse waveform is possible. The terminating resistor R_T is then said to be *matched* to the characteristic impedance of the line. As far as source V is concerned the line might as well be infinitely long.

If $R_T \neq R_0$ all the energy in the forward wave cannot be absorbed by the terminating resistor. The law of conservation of energy demands a reflection phenomenon that is characteristic of all physical systems in which the incident energy cannot be accepted. Thus, light is reflected from the boundary between two media with differing optical properties. Actually, the optical analogy is a good one and it is profitable to regard the termination, in that sense, as a partially silvered mirror.

The reflection of energy, then, causes a reverse wave characterized by $v_r(v_-)$ and $i_r(-i_-)$ which must allow the relationship $R_T = v_T(t_d)/i_T(t_d)$ to be satisfied (Fig. 2.11).

Here, $v_T = v_f + v_r$ $\hspace{6cm}$ (2.24)

and

$\qquad i_T = i_f + i_r$ $\hspace{6cm}$ (2.25)

We now define an important quantity ρ_{VG}, the 'voltage reflection coefficient' at the termination.

$\rho_{VT} \triangleq (v_r/v_f) = (v_-/v_+)$ $\hspace{4.5cm}$ (2.26)

Then, $v_T = v_f(1 + \rho_{VT})$ $\hspace{5cm}$ (2.27)

From eqns (2.19), (2.20), (2.25),

$\qquad i_T = (v_f/R_0) - (v_r/R_0)$ $\hspace{4cm}$ (2.28)

or, $i_T = v_f(1 - \rho_{VT})/R_0$ $\hspace{4.5cm}$ (2.29)

Dividing eqn (2.27) by eqn (2.29) gives,

$\qquad R_T = R_0(1 + \rho_{VT})/(1 - \rho_{VT})$ $\hspace{3.5cm}$ (2.30)

Rearranging eqn (2.30) yields,

$\qquad \rho_{VT} = (R_T - R_0)/(R_T + R_0)$ $\hspace{4cm}$ (2.31)

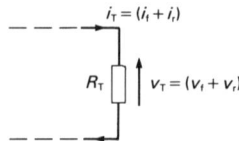

Fig. 2.11 Circuit conditions at termination at $t = t_d$

We can also define a 'current reflection coefficient' at the termination.
Thus,

$$\rho_{IT} \triangleq (i_r/i_f) \tag{2.32}$$

As $i_r = -(v_r/R_0)$ and $i_f = (v_f/R_0)$ it follows that,

$$\rho_{IT} = -(v_r/v_f) = -\rho_{VT} \tag{2.33}$$

For the general case of a terminating impedance Z_G that is not merely
resistive, eqn (2.31) becomes

$$\rho_{VT} = (Z_T - Z_0)/(Z_T + Z_0) \tag{2.34}$$

Equation (2.33) is still valid.

For a terminating impedance that is linear, $Z_T \neq f(i)$, so ρ_{VT}, ρ_{IT} are
constant and independent of signal amplitude. We shall assume that to be the
case, unless otherwise indicated; the case $Z_T = f(i)$ is treated in Chapter 4. As
R_T lies in the range zero (short-circuit termination) to infinity (open-circuit
termination), $+1 \geqslant \rho_{VT} \geqslant -1$.

As it travels backwards along the line, v_r causes the line voltage between the
termination and the sending end to assume a value equal to that at the
termination, $v_T(t_d)$. This can be seen from another viewpoint. During the
course of the reverse wave, by analogy with eqn (2.15),

$$v = -i_r R_0 + \text{constant} \tag{2.35}$$

Substituting $v_r = -i_r R_0$ and 'constant' $= v_f$ gives eqn (2.24). When v_r reaches
the generator, at $t = 2t_d$, there is complete absorption of the energy in the v_r
wave if $R_G = R_0$.

If $R_G \neq R_0$ a reflection occurs at the sending, or generator, end characterized by a voltage reflection coefficient ρ_{VG} and a current reflection
coefficient ρ_{IG}. By reasoning similar to that leading to eqns (2.31) and (2.33),

$$\rho_{VG} = (R_G - R_0)/(R_G + R_0) \tag{2.36}$$

and

$$\rho_{IG} = -\rho_{VG} \tag{2.37}$$

As $\infty \geqslant R_G \geqslant 0$, it follows that $+1 \geqslant \rho_{VG} \geqslant -1$.

For the general case of a generator impedance, Z_G, that is not merely
resistive, eqn (2.36) becomes

$$\rho_{VG} = (Z_G - Z_0)/(Z_G + Z_0) \tag{2.38}$$

Equations (2.37) is still valid.

ρ_{VG}, like ρ_{VT}, is really only of interest provided Z_G is linear, i.e. independent
of signal level.

2.3 MATCHING AND MISMATCHING

If, in Fig. 2.9, Sw is always closed and the battery V is replaced by a generator of an arbitrary waveform, $v_G(t)$, then 'fidelity of signal transmission' requires that $v_T(t)$ be a replica (possibly scaled) of $v_G(t)$ delayed by the transmission time t_d.

For this condition to hold, the line can be impedance-matched at either its sending end or its receiving end or at both ends. Let the scaling factor be λ.

Then, $\lambda \triangleq v_T(t - t_d)/v_G(t)$

The entries in Table 2.1 give λ for those cases in which $v_T(t - t_d)$ is a scaled replica of $v_G(t)$. For cases (i), (iii), (v) the line is matched at the receiving end, i.e. $(R_T/R_0) = 1$; in cases (ii), (iii), (iv), it is matched at the sending end as $(R_G/R_0) = 1$. Whether it is better to match at the sending end or receiving end, or desirable to match at both ends depends on the particular application.

Small departures from the precise matching condition due to resistor tolerances are usually acceptable, as is shown in Example 2.1. This is particularly true if the line is matched at both ends. More serious mismatching can give rise to unwanted multiple reflections (Example 2.2). Gross mismatching, e.g. $(R_T/R_0) = 0$, is often purposely employed in pulse electronics for pulse generation and shaping. Example 2.3 illustrates this.

Example 2.1

A transmission line of 50 Ω characteristic impedance is terminated by a resistive load whose value is 50 Ω ± 5%. Determine the limits of the voltage reflection coefficient at the termination.

Table 2.1 Line matching conditions and resultant scale factor, λ

	Case				
	(i)	(ii)	(iii)	(iv)	(v)
(R_G/R_0)	0	1	1	1	$\neq 1$
(R_T/R_0)	1	∞	1	$\neq 1$	1
λ	1	1	0.5	$\dfrac{R_T}{R_T + R_0}$	$\dfrac{R_0}{R_G + R_0}$

Solution

From eqn (2.31), $\rho_{VT} = (R_T - R_0)/(R_T + R_0)$

But $R_T = R_0 \pm \Delta R$

$\therefore \quad \rho_{VT} = \pm \Delta R/(2R_0 \pm \Delta R)$

As $\Delta R \ll R_0$,

$\rho_{VT} \approx \pm \Delta R/2R_0$

Substituting given data,

$\rho_{VT} = 0 \pm 0.025$

This result has considerable practical significance. Precision 50 Ω terminations for coaxial cable are readily available, but 50 Ω is a non-standard resistor in the preferred ranges. It can be made up by paralleling two 100 Ω resistors, but the voltage reflection coefficient is very small if either of the preferred values 47 Ω or 51 Ω is used.

Example 2.2

Consider the circuit of Fig. 2.9 and assume $R_G = 100$ Ω, $R_0 = 50$ Ω, $R_T = 25$ Ω, $V = 9$ V.

a Sketch and dimension the line conditions for all x for $t = 0+$, $3t_d/4$, $5t_d/4$, $11t_d/4$.
b Plot $v_I(t)$ and $v_T(t)$ for $3t_d > t \geqslant 0+$

Solution

From eqns (2.31) and (2.36), $\rho_{VG} = (100 - 500)/(100 + 50) = +1/3$ and $\rho_{VT} = (25 - 50)/(25 + 50) = -1/3$.

Furthermore, using Fig. 2.10, $v_I(0+) = v_f = (9 \times 50/150)$ V $= 3$ V. The line conditions for $t = 0+$, $t = 3t_d/4$ are shown in Fig. 2.12(a), Fig. 2.12(b).

A reflected wave $v_r = \rho_{VT}v_I = -1$ V is produced at the termination at $t = t_d$. As it travels back, the line voltage is reduced to 2 V. Conditions existing at $t = 5t_d/4$ are shown in Fig. 2.12(c). This is the result when the charged line (Fig. 2.12(d)) is subjected to the reflected wave in Fig. 2.12(e). At the sending end there is a reflected step voltage $\rho_{VT}(-1)$ V $= -1/3$ V. Hence the line conditions at $t = 11t_d/4$ are shown in Fig. 2.12(f), which is a combination of Fig. 2.12(g) and (h).

Composite time pictures for $v_I(t)$, $v_T(t)$ can be built up from sketches such as those of Fig. 2.12. These are shown in Fig. 2.13. Theoretically, reflections continue until $t = \infty$. Then the line plays no part in voltage calculations. As shown in Fig. 2.14, $v_I(\infty) = v_T(\infty) = (25/125)9$ V.

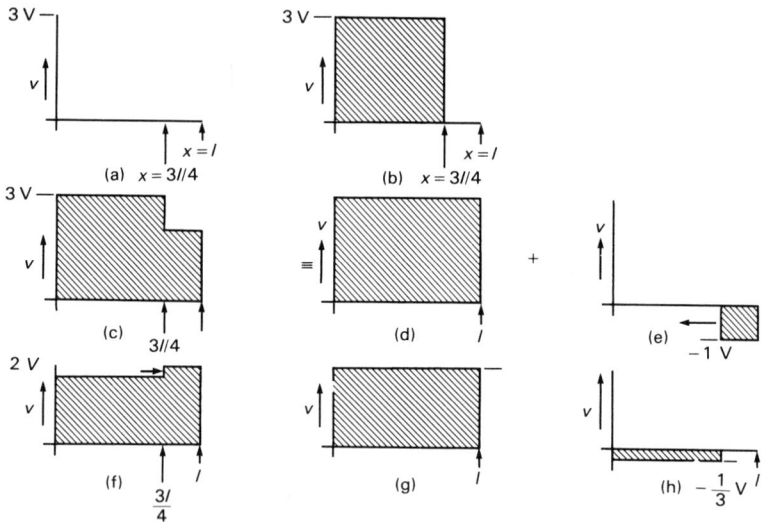

Fig. 2.12   Solution to Example 2.2. Line conditions for all x but specified t. (a) $t=0+$; (b) $t=3t_d/4$; (c) $t=5t_d/4$: a combination of (d) and (e); (f) $t=11t_d/4$: a combination of (g) and (h).

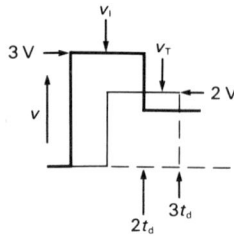

Fig. 2.13   Solution to Example 2.2. $v_I(t)$, $v_T(t)$ for $3t_d > t \geqslant 0+$.

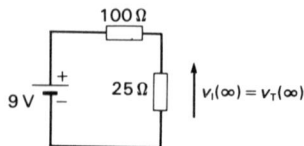

Fig. 2.14   Equivalent circuit for calculating $v_I(\infty) = v_T(\infty)$.

Example 2.3

Consider the system in Fig. 2.9 and assume $R_G = R_0 = 50 \, \Omega$,

$R_T = 0$, $V = 9$ V, $t_d = 10$ ns.

Sketch and dimension $v_l(t)$ for all t, and $v(x)$ for $t_d > t > 0+$.

Solution

Using eqns (2.31) and (2.36), $\rho_{VG} = 0$, $\rho_{VT} = -1$.

By inspection $v_l(0+) = +4.5$ V. When this step wavefront reaches the short-circuit termination there is a reflected wavefront $v_r = -4.5$ V. This follows from the relationship $v_r = \rho_{VT} v_f$, but can also be appreciated physically.

The net voltage at the termination is zero so no energy can be absorbed. As the reflected wavefront travels back along the line, the line voltage is reduced to zero. At the sending end there is no reflection as $\rho_{VG} = 0$. $v_l(t)$ for all t is shown in Fig. 2.15(a), $v(x)$ for $t_d > t > 0$ in Fig. 2.15(b).

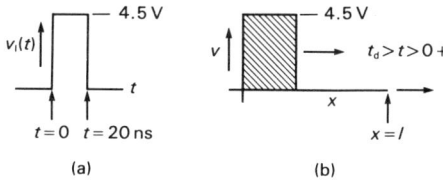

Fig. 2.15 Relating to the solution of Example 2.3.

A short-circuit termination is used to produce a narrow pulse of calculable amplitude and duration from a step-function input signal to the line.

We will now view the reflection phenomenon from a slightly different viewpoint, viz. that using the concept of a 'phantom generator'.

2.4 'PHANTOM GENERATOR' INTERPRETATION OF REFLECTIONS

In section 2.1 we showed that it was possible for two waves to exist on the line at the same time. We can rewrite eqns (2.10) and (2.17(b)), respectively, in the forms,

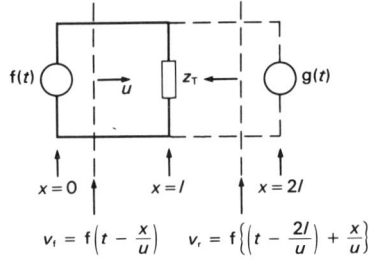

Fig. 2.16 'Phantom generator' approach to line pulse reflections.

$$v_f = f\left(t - \frac{x}{u}\right) \tag{2.39}$$

$$v_r = g\left(t + \frac{x}{u} + \frac{X}{u}\right) \tag{2.40}$$

Let us make the choice $X = -2l$

$$\text{Then,} \quad v_r = g\left\{\left(t - \frac{2l}{u}\right) + \frac{x}{u}\right\} \tag{2.41}$$

Consider, now, Fig. 2.16. $f(t)$ represents a sending-end generator and the forward travelling wave is v_f, given by eqn (2.39). $g(t)$ represents a 'phantom generator' located a distance l beyond the termination, on a line that we imagine extended; this can be seen by putting $x = 2l$ in eqn 2.41. $g(t)$ gives rise to a reverse travelling wave v_r. v_f and v_r travel at the same speed, u, and reach the termination at the same time $t = t_d = (l/u)$.

Then v_f disappears to the right but v_r continues to the left.

$$\text{Thus,} \quad v_T = f\left(t - \frac{l}{u}\right) + g\left(t - \frac{l}{u}\right) \tag{2.42}$$

$$\text{also,} \quad i_T = \frac{1}{Z_0} f\left(t - \frac{l}{u}\right) - \frac{1}{Z_0} g\left(t - \frac{l}{u}\right) \tag{2.43}$$

The minus sign in eqn 2.43 arises for the reasons discussed in section 2.1.

To satisfy the condition $Z_T = (v_T/i_T)$ we must have:

$$Z_T = Z_0\left[1 + \left\{g\left(t - \frac{l}{u}\right) \Big/ f\left(t - \frac{l}{u}\right)\right\}\right] \Big/$$
$$\left[1 - \left\{g\left(t - \frac{l}{u}\right) \Big/ f\left(t - \frac{l}{u}\right)\right\}\right] \tag{2.44}$$

$$\text{But,} \quad \left\{g\left(t - \frac{l}{u}\right) \Big/ f\left(t - \frac{l}{u}\right)\right\} = g(t)/f(t) \tag{2.45}$$

because v_f, v_r travel with the same velocity.

If we make the substitution,

$$\rho_{VT} = g(t)/f(t) \tag{2.46}$$

it follows from eqns (2.44), (2.45) and (2.46) that,

$$\rho_{VT} = (Z_T - Z_0)/(Z_T + Z_0) \tag{2.47}$$

This is, of course, eqn (2.34), presented earlier.

If Z_T, Z_0 are resistive, only, in nature ρ_{VT} is a real number.

The phantom generator $g(t)$ is a scaled down, and possibly inverted form of $f(t)$ and disappears at $t = 2t_d$. R_G is missing from Fig. 2.16 as $f(t)$ is taken to be the line input. If $R_G \neq R_0$ any reflection at the sending end causes the appearance of a new phantom generator in place of $g(t)$. (A reflection at the generator end can also be imagined in terms of a phantom generator $f_1(t)$, say, located at $x = -l$ and sending a wave in the forward direction).

For the particular case considered in Example 2.2, $f(t)$ is a 3 V positive-going step occurring at $t = 0$. Hence, $g(t)$ appears to be a -1 V step for the first reverse waveform down the line because $\rho_{VT} = -1/3$ and eqn (2.46) is applicable. Thus Fig. 2.16 reduces to Fig. 2.17 for $2t_d \geqslant t > 0+$.

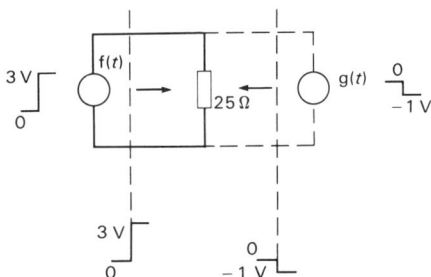

Fig. 2.17   Relating to Fig. 2.12(a) to (e).

The concept of a phantom generator is unnecessary for the understanding of wave transmission when step-function signals only are considered. However, it does help in determining the time-function of line voltage at a termination when waveforms of arbitrary shape are used, as will be seen in the next chapter.

PROBLEMS

P2.1   Using simple dimensional analysis show that,

   a $[\sqrt{(L/C)}] = [R]$
   b $[\sqrt{(LC)}]$ $= [T]$, where $T = $ time/unit-length

P2.2 A manufacturer specifies a certain cable (type RG8A/U) as having $C = 29.5$ pF/ft and $L = 73.75$ nH/ft. Assumming the line is lossless, calculate

    a Characteristic impedance, Z_0;

    b Time-delay per foot length;

    c Time-delay per metre length.

P2.3 A loss-free transmission line has $Z_0 = 75$ Ω, $C = 69$ pF/m. Calculate the time-delay per metre.

P2.4 Show, by partial differentiation, that the following expressions satisfy eqn (2.8). (In **c**, X is a constant).

    a $v = f\{t - x\sqrt{(LC)}\}$

    b $v = g\{t + x\sqrt{(LC)}\}$

    c $v = g\{t + x\sqrt{(LC)} + X\sqrt{(LC)}\}$

    d $v = f\{t - x\sqrt{(LC)}\} + g\{t + x\sqrt{(LC)}\}$

P2.5 A positive-going step-voltage of magnitude V propagates down an initially uncharged, semi-infinite, loss-free line of characteristic resistance R_0.

    a Show that the energy, W_E, stored in the electric field when the waveform has reached a point x' along the line, is: $W_E = (Cx')V^2/2$.

    b Show that the energy, W_M, stored in the magnetic field is: $W_M = (Lx')I^2/2$, where $I = V/R_0$

    c Hence, show that $W_E = W_M$

    d Show that the rate at which energy is increasing (i.e. power is continuously supplied to the line), is VI.

P2.6 By arguments similar to those leading to eqns (2.4), (2.6) show that for a line with loss,

$$-\left(\frac{\partial v}{\partial x}\right) = Ri + L\left(\frac{\partial i}{\partial t}\right)$$

$$-\left(\frac{\partial i}{\partial x}\right) = Gv + C\left(\frac{\partial v}{\partial t}\right)$$

P2.7 Show, from P2.6, that

$$\left(\frac{\partial^2 v}{\partial x^2}\right) = LC\left(\frac{\partial^2 v}{\partial x^2}\right) + (RC + LG)\left(\frac{\partial v}{\partial t}\right) + RGv$$

P2.8 If, in P2.7, $RC = GL$ we have the case of a 'distortionless line'. This is one in which there is a change in scale due to attenuation, but not shape, as a waveform is propagated along it.

    Show, by partial differentiation, that the following expression is applicable to a distortionless line:

$$v = \exp(-\alpha x)f\left(t - \frac{x}{u}\right)$$

where $\alpha = R\sqrt{\dfrac{C}{L}}$ and $u = 1/\sqrt{(LC)}$

3 The Reflection Chart

The 'book-keeping' exercise associated with a problem involving multiple reflections is simplified by the use of a '*reflection chart*' or '*lattice-diagram*', which is a convenient two-dimensional space–time plot showing wavefront progress on a line.

3.1 THE REFLECTION CHART: STEP-INPUT VOLTAGE

The reflection chart and its value in solving line problems will first be explained by reference to the particular case of a step-voltage input waveform applied to an initially uncharged line. Fig. 3.1 shows a reflection chart. By

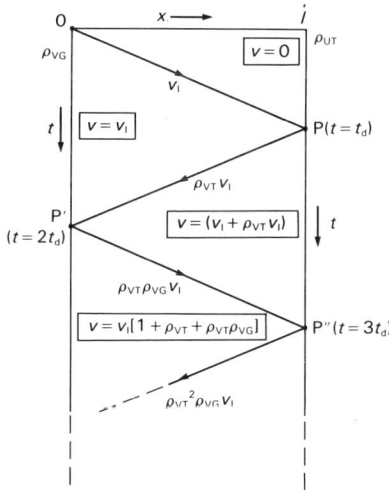

Fig. 3.1 'Reflection chart' or 'lattice diagram' for line voltage reflections. A 'boxed' value for v is the sum of the individual waves above that region and represents the net voltage there.

convention, distance along the line increases horizontally to the right, from $x = 0$ to $x = l$, and time increases vertically downwards. The reflection coefficients ρ_{VG}, ρ_{VT} are shown at $x = 0$, $x = l$.

The first forward wave due to the initial step is indicated by the arrowed line 0P. The attached symbol, v_I, gives the step-voltage in magnitude and sign. The magnitude of the slope of 0P is the signal velocity u, so clearly the diagram is not drawn to scale; it is schematic only. 0P has a negative slope only because of the arbitrary, but convenient, choice of t increasing downwards.

The progress of a reflected wave is shown by the line PP' with an arrow indicating motion back towards the input end of the line, and a symbol $\rho_{VT} v_I$ giving the magnitude and sign of the reflected voltage step. The slope of PP' is equal in magnitude, but opposite in sign, to that of 0P. Subsequent reflected voltages are shown by a continuing zig-zag line with appropriately labelled segments. The zig-zag line stops at $t = t_d$ if $\rho_{VT} = 0$, or at $t = 3t_d$ if $\rho_{VT} \neq 0$ but $\rho_{VG} = 0$.

For the first forward wave $v(x, t) = 0$ for $t < (x/u)$. Thus for all pairs of co-ordinates x, t that fall within the area 0/P (but not *on* 0P itself), $v = 0$ and to remind us of this fact the value is shown boxed. Similarly for all pairs of co-ordinates that lie in the area 0PP' (but not *on* PP' itself) the line voltage is v_I, which is also shown boxed. Generally, a boxed value for v in a space is the sum of the individual waves above that region and represents the net voltage in the space considered.

To find $v(x)$ for a specified $t = t'$, consider Fig. 3.2(a). We draw a horizontal line aa' at $t = t'$. In this case $t_d > t' > 0+$: aa' cuts the zig-zag line on

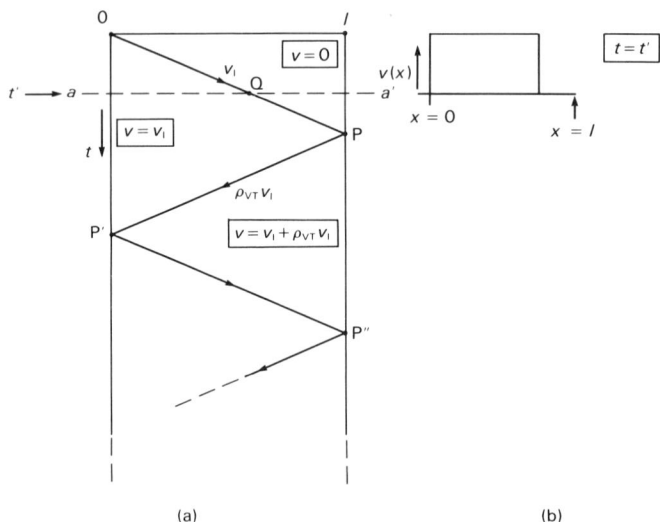

Fig. 3.2 Finding $v(x)$ at a given t using the reflection-chart.

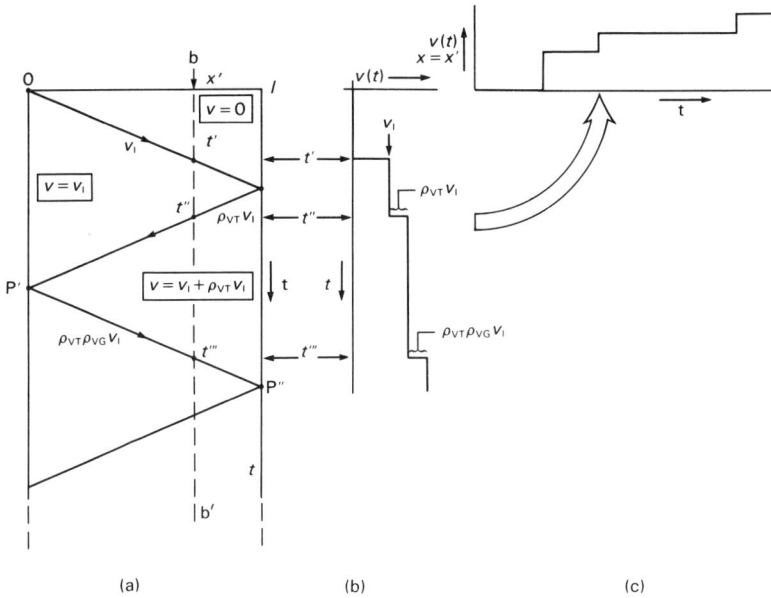

Fig. 3.3   Finding $v(t)$ at a given x.

the section 0P at the point Q corresponding to $x = x'$. Q defines the point reached by the wavefront at t'. That part of aa' lying within the triangle 0PP' corresponds to the part $(x' \geqslant x > 0)$ that is charged up to the potential shown boxed, v_1. Similarly, that part of aa' lying within the triangle 0lP corresponds to the section $(l > x > x')$ of the line that is uncharged, as shown by the boxed value. Fig. 3.2(b), showing $v(x)$ for $t = t'$, is quickly obtained by horizontal projection.

To find $v(t)$ for a specified $x = x'$, consider Fig. 3.3(a). We draw a vertical line bb' at $x = x'$. This cuts the zig-zag line at t', t'', etc. Fig. 3.3(b), showing $v(t)$ for $x = x'$, is easily obtained from Fig. 3.3(a) by horizontal projection. Conventionally Fig. 3.3(b) is rotated through 90° to give the plot in Fig. 3.3(c).

Before working two examples, note that we have discussed the reflection chart for voltage waves. This is because voltage waveforms are, in principle, observable with an oscilloscope. A reflection chart for line currents can equally well be drawn (see Problems P3.1, P3.2).

Example 3.1

In Fig. 3.4, Sw closes at $t = 0$ and remains closed. Draw a reflection chart and hence sketch and dimension,

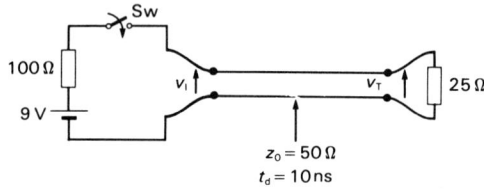

Fig. 3.4   Circuit for Example 3.1.

a $v(l/2, t)$ for 30 ns $\geqslant t \geqslant 0+$.
b $v(x)$ for $t = 22.5$ ns.

Solution

Fig. 3.5 follows directly from the application of the procedures outlined above.

The case of an initially charged line lends itself quite simply to a reflection chart approach as is shown in the next example.

Example 3.2

In Fig. 3.6, Sw closes at $t = 0$ and remains closed. Draw a reflection chart and sketch and dimension,

$$v_I(t) = v(0, t) \text{ and } v_T(t) = v(l, t)$$

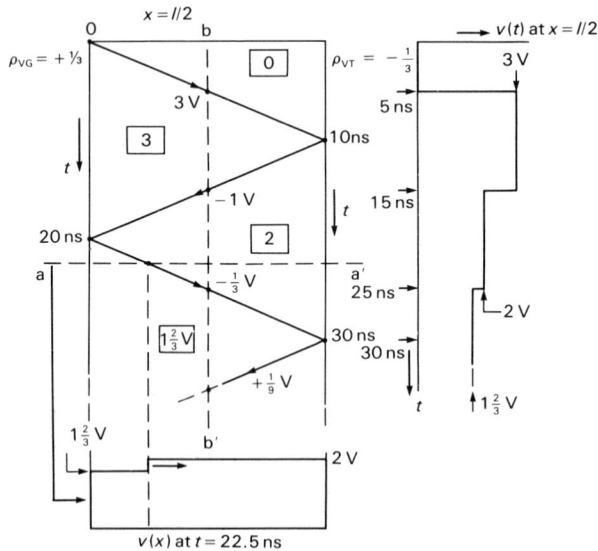

Fig. 3.5   Solution to Example 3.1.

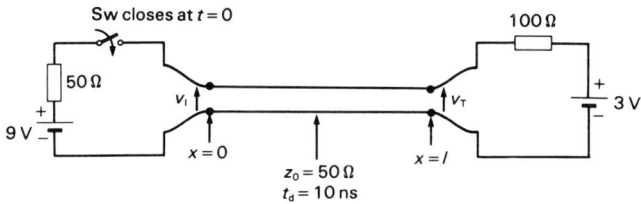

Fig. 3.6   Circuit for Example 3.2.

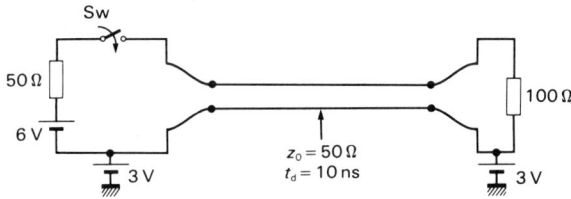

(a)

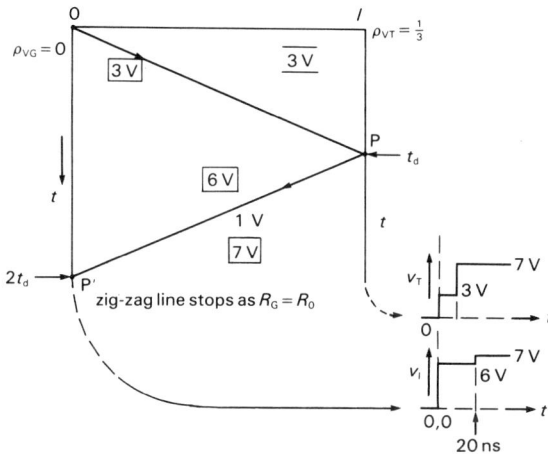

(b)

Fig. 3.7   Solution to Example 3.2.

Solution

For calculation purposes Fig. 3.7(a) is equivalent to Fig. 3.6. By relocating the voltage source at the termination and modifying the source directly switched to the line at $t = 0$, the problem is reduced to one involving voltage *changes* added to a constant value.

At $t = 0$ the voltage change at the input to the line is 3 V because 6 V is divided equally between the source resistance and the characteristic resistance of the line. The first forward voltage wave is thus a $+3$ V step. This is reflected at $x = l$ where $\rho_{VT} = 1/3$. The reverse voltage step, $+1$ V, is absorbed at the sending end at $t = 2t_d$ because $R_G = Z_0 = R_0$. Fig. 3.7(b) shows all relevant information (note that triangle 0/P has a boxed voltage 3 V corresponding to the initial voltage on the line) and required sketches of $v_I(t)$, $v_T(t)$.

It is interesting to note that $v_I(0+)$ can be calculated directly from the equivalent circuit of Fig. 3.8. The charged line at $t = 0+$ appears as a 3 V source in series with its 50 Ω characteristic resistance, *not* 3 V in series with 100 Ω. (That condition holds only when all reflections have ceased.) The reason for this stems from eqn (2.15); the constant term in that equation is 3 V.

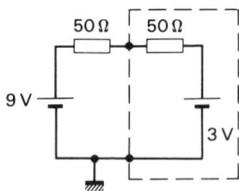

Fig. 3.8   Circuit for calculating initial line current in Example 3.2.

Further physical insight into line operation may be obtained by extending the reflection chart approach to a third dimension so that line voltage $v(x, t)$ is plotted vertically.

A solid with a 'stepped' surface is generated by the wavefront reflections. From this we can visualize $v(x)$ at a given t and $v(t)$ for a given x. Thus in Fig. 3.9, $v(x)$ at $2t_d > t \, (= t') > t_d$ is the profile of the cross-section of the solid as defined by its intersection with the perpendicular plane $t = t'$. For ease of drawing, the case $\rho_{VT} > 0$ is considered. Similarly, $v(t)$, for a given $l > x(= x') > 0$, is the profile of the cross-section of the solid as defined by its intersection with the perpendicular plane $x = x'$.

3.2   RECTANGULAR PULSE DRIVE

A rectangular pulse can be considered as comprising two equal amplitude steps, of opposite polarity, with a time-delay equal to the pulse duration. The principle of superposition applies for linear source and terminating impedances, so we can find the line voltage by summing, algebraically the effects of each step taken separately. We choose as an example a pulse whose duration is less than the two-way delay time on the line but the principle is generally applicable.

For the circuit of Fig. 3.10, v_G is a 9 V, 2 ns pulse applied from a 25 Ω

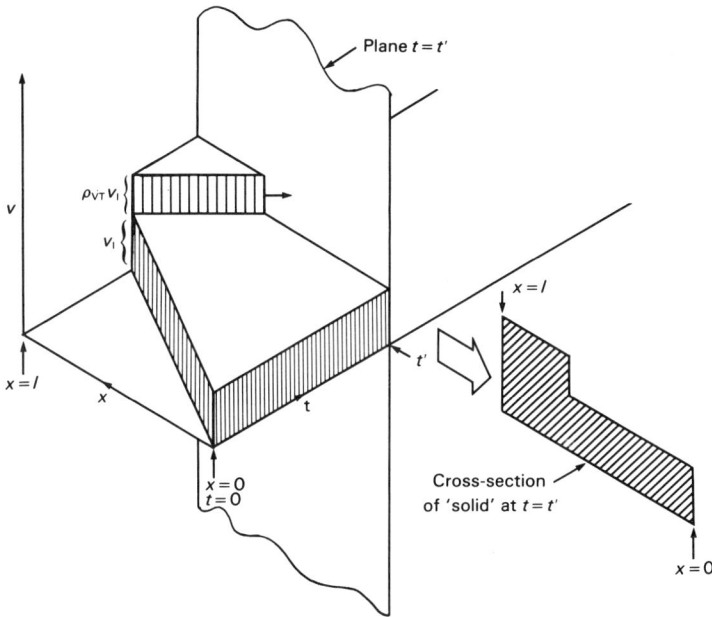

Fig. 3.9 Extension of Reflection Chart to three dimensions.

source resistance to an uncharged line having $R_0 = 50\ \Omega$, $t_d = 10$ ns. $v_I(0+) = 6$ V, and this forward wavefront is followed 2 ns later by a -6 V step as shown in Fig. 3.10(c). The reflection chart, Fig. 3.10(d), includes small triangular areas in which the line voltage is the algebraic sum of the forward and reflected wavefronts. The waveforms for $v(0,t) = v_I(t)$ are shown in Fig. 3.10(e).

3.3   INPUT WAVEFORMS WITH FINITE RISE- AND FALL-TIMES

The three-dimensional model introduced in section 3.1 is useful in clarifying the understanding of instantaneous line conditions when an input waveform with non-zero rise-time is applied. Thus, in Fig. 3.11(a) a truncated-ramp function voltage waveform is applied to the input of a line. $v(x)$ for the particular case $t_d > t(= t') > 0$, obtained from a cross-section at $t = t'$, is shown in Fig. 3.11(b). (Compare with Fig. 2.7 and the discussion leading to it in section 2.2.)

Now that we can sketch $v(x)$ for a given t' ($<t_d$) we will use the 'phantom generator' approach to section 2.4 to obtain pictures of $v(x)$, at chosen times,

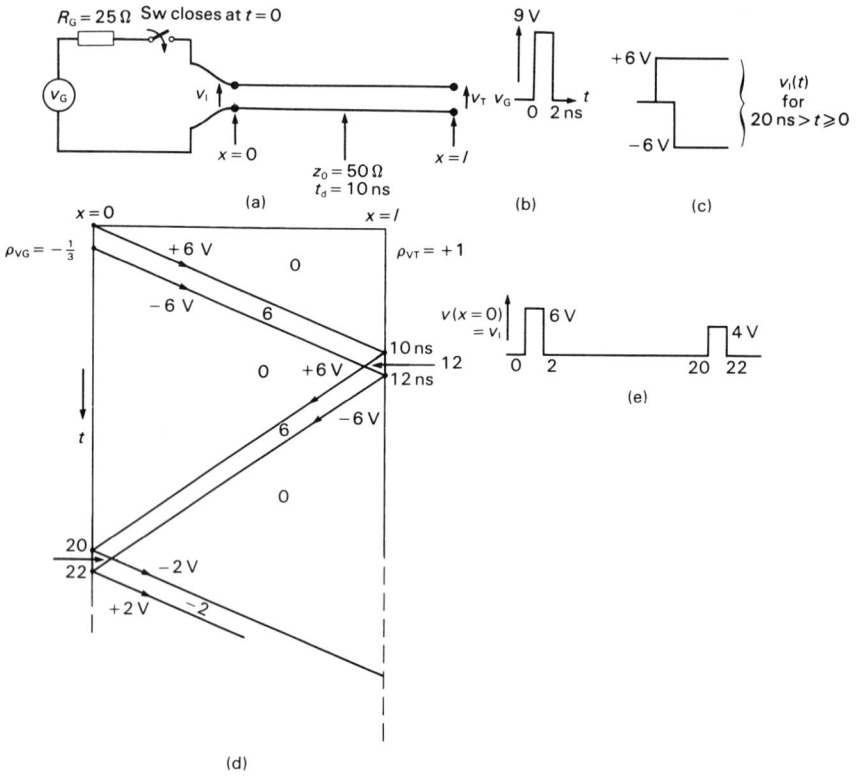

Fig. 3.10 Reflections with a rectangular pulse input. (a) Circuit; (b) $v_G(t)$; (c) Equivalent form for (b); (d) Reflection chart; (e) $v_I(t)$ for 40 ns > t > 0+.

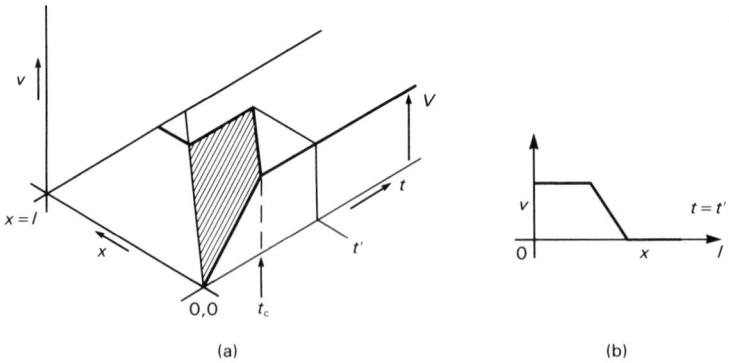

Fig. 3.11 Showing progress of a truncated-ramp waveform on a line.

after a reflection has occurred at both $x = l$ and $x = 0$. We will then see how a normal reflection chart, suitably interpreted, can be used as a short-cut method for obtaining $v_I(t)$, $v_T(t)$ for applied waveforms of arbitrary shape and duration.

Consider Fig. 3.12. In (a), a waveform v_G is applied through a 25 Ω resistor to an open-ended and initially uncharged line having $Z_0 = 50$ Ω, $t_d = 16$ ns. v_G, as specified in (b) is a truncated-ramp voltage waveform with a transition time, t_c, of 8 ns and an amplitude 9 V. To determine the progress of the wave-form along the line we follow the time-sequence of sketches for a forward wave v_f and any existing reverse wave(s) v_r.

In (c) are shown line conditions due to the first forward wave v_f, at con-veniently chosen values of t (in ns). The dotted sections serve to indicate what would happen if the line were extended. After $t = 24$ ns, v_{f1} disappears forever to the right.

The line conditions that result solely from the existence of the first reverse wave v_{r1} are shown in (d); v_{r1} is considered to be produced by a 'phantom generator' (located at $x = 2l$) and is a mirror image of v_{f1} because $\rho_{VT} = +1$. v_{r1} does not affect the line until $t \geqslant 16$ ns. From (c) and (d) we can determine $v(x)$ at a given t for 32 ns $> t > 0$ ns., or $v(t)$ at a given x for the same time interval. Thus (e) shows $v(x)$ at $t = 26$ ns and is obtained by algebraic addition of the ordinates of v_{f1} and v_{r1} at the time $t = 26$ ns.

Similarly, by adding v_{f1} and v_{r1} at $x = l$ at the given circled times, we obtain $v(l, t) = v_T(t)$, shown in (f). When v_{r1} reaches the input to the line at $t = 32$ ns there is a reflected waveform v_{f2}. (This can be imagined to arrive from a phantom generator at $x = -l$). v_{f2} is a scaled-down and inverted version of v_{r1}, as $\rho_{VG} = -1/3$, see (g).

As v_{f2} progresses down the line it adds algebraically to the existing voltage. Thus, (h) shows line conditions at $t = 40$ ns and $t = 42$ ns. When $t = 48$ ns, v_{f2} reaches $x = l$, and there is another reflection v_{r2}. We can carry on sketching pictures of line conditions but will not do so because no new principle is involved.

This example is illustrated in some detail for explanation purposes. Quite often interest centres only on $v_I(t)$ and $v_T(t)$, and to obtain these it is certainly not desirable to follow the somewhat tedious procedure outlined, however informative it may be, if it can be avoided. A short-cut method is available, and easily applied, if $t_c < 2t_d$.

Thus, we imagine the input to be a *step* and use a reflection chart as in (i). From this we easily obtain $v_I(t)$ as shown in (j). We now change the step edges to scaled replicas — due account being given to any polarity inversion accompanying a reflection — of the actual generator waveform edges. In the case considered, and shown in (k), this means truncated-ramps. The limiting case for finding $v_I(t)$, $v_T(t)$, simply, using this method occurs when $t_c = 2t_d$. Suppose that in Fig. 3.12(a), $t_d = 4$ ns but all else remains unchanged. Fig. 3.13 shows the resulting $v_I(t)$ for 24 ns $\geqslant t > 0$. Comparing this with

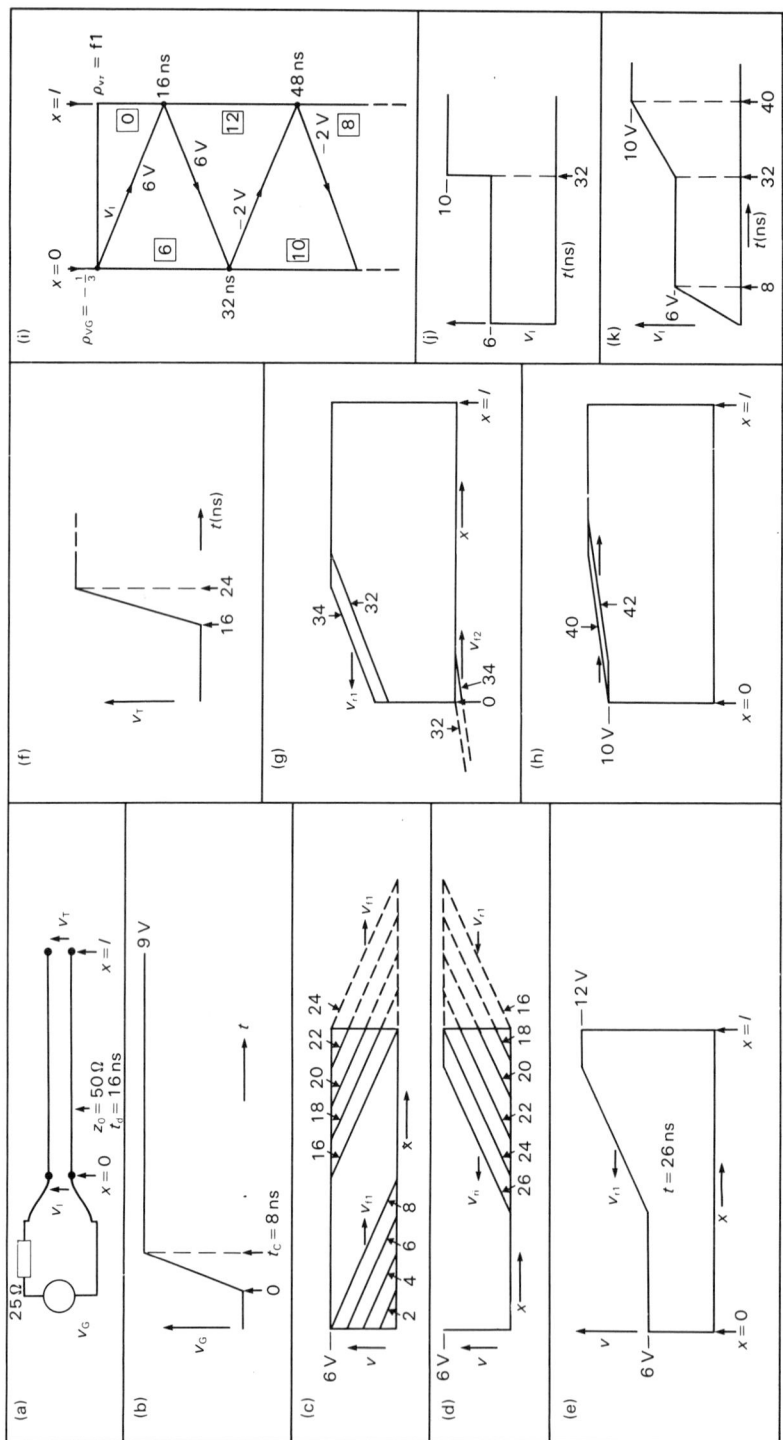

Fig. 3.12 Truncated-ramp voltage drive. (a) Circuit; (b) $v_G = f(t)$; (c) $v = f(x)$ for indicated times, (ns), for forward wave v_{f1}; (d) $v = f(x)$ for indicated times for reverse wave v_{r1}; (e) $v = f(x)$ for $t = 26$ ns, obtained by summing respective parts of (c) and (d); (f) $v_T(t)$ for 48 ns> $t>0$; (g) v_{r1} and v_{f2} at $t = 32$ ns and $t = 34$ ns; (h) $v = f(x)$ for indicated times; (i) Reflection chart for *step* voltage input; (j) $v_1(t)$, for 48 ns> $t>0$ ns, for a step input; (k) $v_1(t)$, for truncated-ramp voltage input.

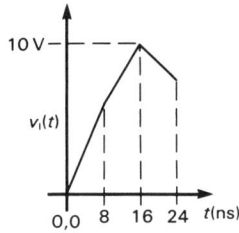

Fig. 3.13 $v_I(t)$ for Fig. 3.12(a), when $t_d = 4$ ns. Compare with Fig. 3.12(k).

Fig. 3.12(k) we note that the 'flat' sections of that waveform have now disappeared.

Figure 3.13 shows a waveform with 'overshoots'. At $t = 16$ ns, $v_I = 10$ V and at $t = 24$ ns, $v_I = 8.67$ V. The final value for v_I (as $t \to \infty$) is 9 V. If we imagine v_G in Fig. 3.12(a) to have 'rounded corners', as in Fig. 3.14(a), then the waveform for $v_I(t)$ is also a smooth curve, Fig. 3.14(b). A 'ringing' waveform of this type is commonly observed, in practice, when a fast pulse edge is examined by an oscilloscope whose incorrectly terminated leads introduce a time-delay comparable with the edge rise-time.

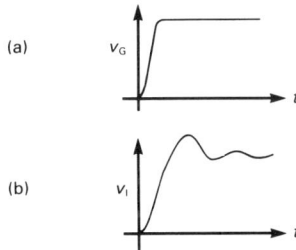

Fig. 3.14 (a) Truncated-ramp drive waveform ($t_c = 2t_d$) with 'rounded corners' applied to circuit of Fig. 3.12(a); (b) Resultant 'ringing' waveform.

The final case mentioned here concerns the condition $t_c \gg t_d$. For $R_G = R_0$, $v_I(t)$ and $v_T(t)$ are obtained easily by the addition of two waveforms (see Problem P3.11). However, if $R_G \neq R_0$ one way of dealing with the problem is to 'slice' the input waveform into a series of sections each $2t_d$ in duration and to treat each separately. The Principle of Superposition applies so the contributions to $v_I(t)$ because each section taken separately can be summed. Thus in Fig. 3.15, for the particular case $t_c = 4t_d$, v_G may be regarded as comprising the components v_{G1}, v_{G2}.

It must be remembered that for $t_c \gg t_d$ the system ceases to be 'distributed', as was pointed out in Chapter 1.

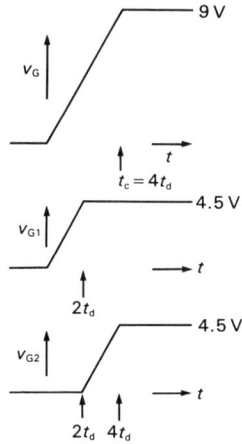

Fig. 3.15 Resolution of v_G into two components each with edges $2t_d$ in duration for the case $t_c = 4t_d$.

PROBLEMS

(Note: 'Sketch' implies 'dimensioning'.)

P3.1 Refer to Fig. P3.1. Let $V = 1$ V, $R_G = 50$ Ω, $R_T = 0$.

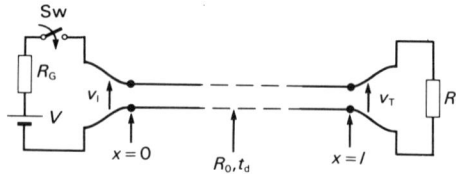

Fig. P3.1 Refers to P3.1 to P3.4 inclusive, and also P3.8. $t_d = 10$ ns, $R_0 = 50$ Ω. In all cases Sw closes at $t = 0$ and remains closed, unless otherwise stated in the question.

    a Draw voltage and current reflection charts for 20 ns $\geqslant t \geqslant 0+$.
    b Sketch $v_I(t)$, $v_T(t)$ for this time interval.
    c Sketch $i(x)$ for $t = 15$ ns.

P3.2 Refer to Fig. P3.1. Let $V = 1$ V, $R_G = 50$ Ω, $R_T = \infty$.
    a Draw voltage and current reflection charts for 25 ns $\geqslant t \geqslant 0$.
    b Sketch $v_I(t)$, $v_T(t)$ for this time interval.
    c Sketch $i(t)$ at $x = l/2$.

P3.3 Refer to Fig. P3.1. Let $V = 3$ V, $R_G = 25$ Ω, $R_T = 100$ Ω.

    a Sketch and dimension $v_I(t)$, $v_T(t)$ for 35 ns $\geqslant t \geqslant 0+$.
    b Sketch $v(t)$ at $x = l/4$ for 35 ns $\geqslant t \geqslant 0+$.
    c Sketch $v(x)$ for $t = 25$ ns.

P3.4 Refer to Fig. P3.1. Let $V = 1$ V, $R_G = 100$ Ω, $R_T = 50$ Ω. Sw closes then opens at $t = 10$ ns. Sketch $v_I(t)$, $v_T(t)$ for 25 ns $\geqslant t \geqslant 0+$.

P3.5 Refer to Fig. P3.5. For $x = l/2$, sketch $v(t)$ for 30 ns $\geqslant t \geqslant 0+$.

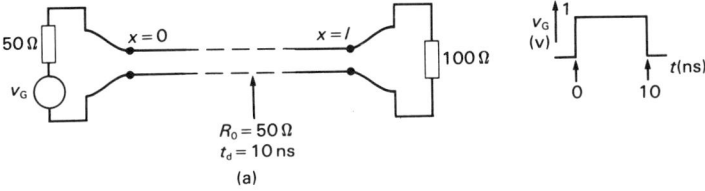

Fig. P3.5   Relating to problem P3.5.

P3.6 Refer to Fig. P3.6. Sketch and explain, briefly, the shape of $v_I(t)$ for 70 ns $\geqslant t \geqslant 0$ ns.

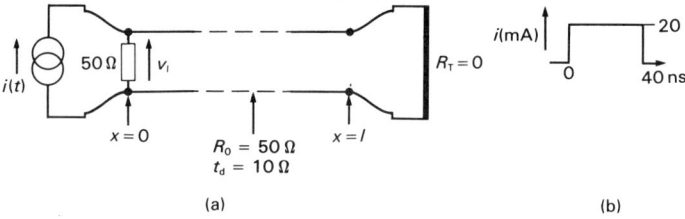

Fig. P3.6   Input current waveform $i(t)$ for (a) is shown in (b).

P3.7 Consider Fig. 3.6 of the text. Assume Sw *opens* at $t = 40$ ns. Sketch $v_I(t)$, $v_T(t)$ for 65 ns $\geqslant t \geqslant 40$ ns.

P3.8 Refer to Fig. P3.1. Let $R_T = \infty$. Show that:
    a $\Delta v_I(2t_d) = \{R_0/(R_0 + R_G)\} \{1 + \rho_{VG}\} V$ (Δ indicates 'change in')
    b For $t = r(2t_d)$, where $r = 2, 3, 4 \dots$.
    $[\Delta v_I\{r(2t_d)\}/\Delta v_I\{(r-1)(2t_d)\}] = \rho_{VG}$

P3.9 From P3.8, by summing the geometrical progression resulting from the increments in v_I, show that for $R_G \gg R_0$

$$v_I(t) \approx \{R_G/(R_G + R_0)\} \{1 - \rho_{VG}^n\} V \approx V(1 - \rho_{VG}^n)$$

where $t = n(2t_d)$, and $n \geqslant 2$.

P3.10 In P3.9, let $\rho_{VG}^n = \exp(-t/\tau)$. Then, taking logarithms and using the binomial expansion for the case $R_G \gg R_0$, show that
    a $\tau \approx R_G C_T$, where $C_T = $ *total* line capacitance;
    b $v_I(t) \approx V[1 - \exp(-t/C_T R_G)]$.
    Give a physical interpretation of **b**.

P3.11   Refer to Fig. P3.11. Sketch $v_I(t)$, $v_T(t)$ for these cases:

    a $R_T = \infty$, $T = 6$ ns;   **c** $R_T = 0$, $T = 6$ ns;

    b $R_T = \infty$, $T = 100$ ns;   **d** $R_T = 0$, $T = 100$ ns.

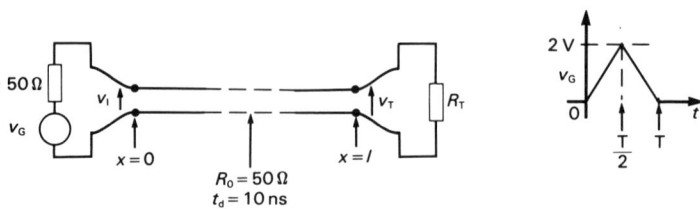

Fig. P3.11   Relating to P3.11.

4 'Sliding-Load-Line' Analysis of Pulses on Lines

In Chapters 2 and 3, we considered line pulse behaviour with 'linear' terminations and to concentrate on the subject under discussion we glossed over the precise meaning of the word 'linear'. For an ideal resistor the property of 'linearity' means that (v/i) is constant, v and i being, respectively, the potential difference across the resistor and the current flow in it. The ratio is, of course, the resistance R and constancy means independence of signal level, so R is not dependent on the magnitude of v.

For a capacitor, 'linearity' means that (q/v) is constant, q being the stored charge and v the applied voltage. The proportionality factor is the capacitance C, and the constancy of C means that it is independent of the magnitude of q.

Now, as $i = (dq/dt)$ and C is constant, it follows that $i = C(dv/dt)$. So, for a capacitor there is a linear relationship between current, i, and the rate of change of v.

In electrical engineering, generally, 'linearity' means that two or more variables are related either by constant factors of proportionality or by differential equations with constant coefficients. For a linear system the Principle of Superposition applies: thus, if an input $v_{I1}(t)$ produces an output $v_{01}(t)$ and an input $v_{I2}(t)$ produces an output $v_{02}(t)$ then an input $\{v_{I1}(t) + v_{I2}(t)\}$ produces an output $\{v_{01}(t) + v_{02}(t)\}$.

As was pointed out in Chapter 2, eqns (2.8) and (2.9) for wave motion on a loss-free line are linear. If linear terminating impedances are used, the whole system (line + terminating impedances) is linear. The reflection chart, which incorporates the principle of superposition, is applicable and capable of supplying all the information normally required in practical circuits in which line pulse reflections occur.

Suppose, however, that a line is fed by a voltage generator whose source impedance Z_G is linear and is terminated by a load impedance Z_T that is non-linear. In particular we can imagine the load to be a semiconductor diode. This is definitely nonlinear because, under d.c. conditions, the current in the diode is exponentially related to the applied voltage.

The fundamental implication of nonlinearity in Z_T is that ρ_{VT}, ρ_{IT} are voltage-dependent and are thus not constant with signal level. Thus, a reflection chart approach — with all its appealing elegance — is not directly applicable.

To deal with practical nonlinear impedances, particularly those that are met with at the outputs and inputs of digital logic circuits, a new approach is required. A convenient technique employs what we will term the concept of the 'sliding-load-line' (SLL). The reason for this description will become clear as we proceed.

For comparison it must be remembered that the usual reflection chart is a pictorial representation — normally of v v. t — that is only schematic and does not require accurate plotting. Normally it involves *changes* in line conditions. The SLL approach is graphical and requires accurate plots, to scale, in cartesian co-ordinates of v v. i. It always involves *total* line voltages or currents.

It is possible to *use* the SLL approach on a purely 'cook-book' basis without understanding its physical justification. However, the author has found that this procedure is unsatisfactory for students with questioning minds so the SLL approach is developed from physical reasoning in the next section. To smooth the transition from familiar to less familiar concepts we again solve a problem, involving linear terminations, that was dealt with in Chapter 3. We then extend the approach to nonlinear terminations.

4.1 RESISTIVE TERMINATIONS

Figure 4.1 refers to Example 2.2. When Sw closes, a 'snapshot' of initial line conditions at the input to the line can be obtained from the graphical

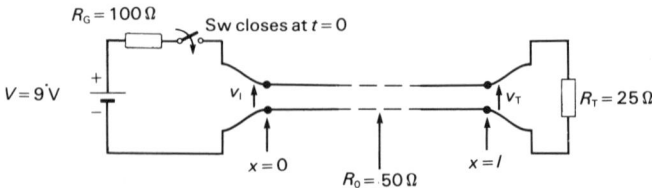

Fig. 4.1 System of Example 2.2, redrawn for ease of reference.

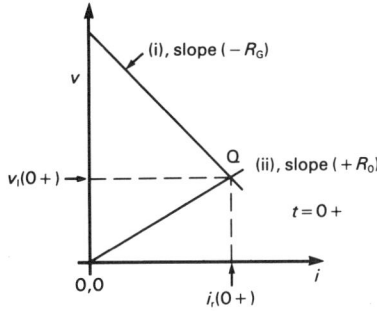

Fig. 4.2 Graphical determination of $v_I(0+)$.

construction in Fig. 4.2 (we have chosen to plot $v = f(i)$, but the alternative choice $i = f(v)$ is equally valid).

In Fig. 4.2 curves (i) and (ii) are both straight lines: (i), with slope $-R_G$, represents the characteristic due to the battery V and source resistance R_G and is defined by $v_I(0+) = V - R_G i_I(0+)$; (ii), representing the instantaneous relationship between $v_I(t)$ and $i_I(t)$ at $t = 0+$ for the uncharged line, has a slope $+R_0$ and passes through the origin because $v_I(0+) = R_0 i_I(0+)$. Note that capital subscripts are used for v and i because we are concerned with total instantaneous values. The intersection point, Q, gives $v_I(0+) = VR_0/(R_0 + R_G)$, the same as that indicated in Fig. 2.10.

Suppose we now wish to represent line behaviour at point x' at a time t' ($= x'/u$) by an instantaneous time plot similar to that of Fig. 4.2. Such a plot is shown, correspondingly labelled, in Fig. 4.3. The dynamic resistance seen looking in *either* direction from the selected point is the characteristic resistance R_0. Thus curve (ii) represents, as before, the line to the right of x' while (i) with a slope $-R_0$ (rather than $-R_G$) now represents the resistance looking backwards towards the source. The point of intersection, Q', of (i) and (ii) is the line voltage; but this is equal to $v_I(0+)$ because in travelling down the line the wave v_f causes successive points to assume this value. The

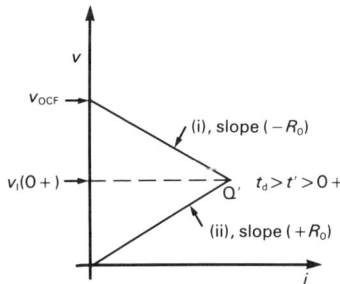

Fig. 4.3 Construction giving line behaviour at point x' reached by forward wavefront at t', where $t_d > t' > 0+$.

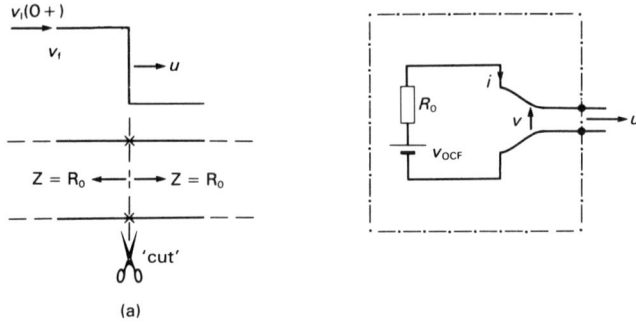

(a)

Fig. 4.4   Illustrating the sliding Thévenin source or sliding-load-line concept for the forward wave.

apparent battery voltage seen from x' is thus $v_{OCF} = 2v_I(0+)$. The letters OC refer to open-circuit.

As far as points to the right of x' are concerned we can imagine the line to be cut at x' and the section to the left to be replaced by a Thévenin source circuit that 'slides' along the line, i.e. we have a 'sliding-load-line'. This process is illustrated in Fig. 4.4; the equivalent source components in the chain-dotted rectangle move to the right with the velocity u of the v_F wavefront. If we 'stack-up', one behind the other, a series of plots such as Fig. 4.2 and Fig. 4.3 we obtain the three-dimensional v, i, t plot in Fig. 4.5, and this provides a good visualization of wavefront progress. The projection on a plane drawn perpendicular to the i axis and passing through Q' gives $v = f(t)$,

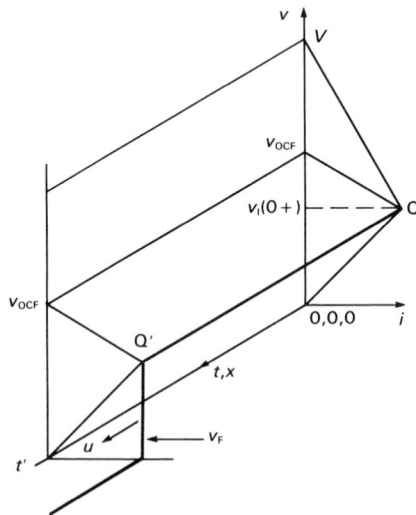

Fig. 4.5   A v, i, t plot of line behaviour for $t_d > t > 0+$.

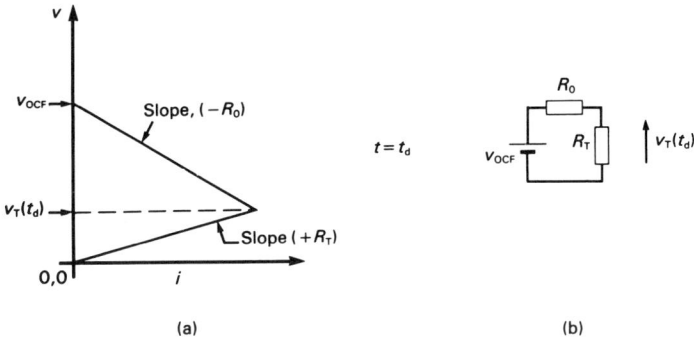

Fig. 4.6   Line conditions at $t = t_d$. (a) Graphical construction;
(b) Equivalent circuit.

whereas the projection on a plane drawn perpendicular to the v axis yields $i = f(t)$.

The conditions at $x = l$, at $t = t_d$, are found by considering the v, i curve there — see Fig. 4.6(a): $v_T(t_d)$ can be found from this or the associated circuit diagram in Fig. 4.6(b).

By inspection, $v_T(t_d) = v_{OCF} R_T/(R_T + R_0)$ (4.1)

or, $v_T(t_d) = 2v_I(0+) R_T/(R_T + R_0)$ (4.2)

Using the numerical data for our example gives: $v_I(0+) = +3$ V; $v_{OCF} = 6$ V; $v_T(t_d) = 2$ V. The value of $v_T(t_d)$ agrees, of course, with that which can be obtained from the reflection chart in Fig. 3.5.

As $R_T \neq R_0$ there is a reflected wave v_R which proceeds from R_T to the battery and causes successive points on the line to take on the voltage value $v_T(t_d)$ as it reaches them.

The line voltage at any point due to v_R is given by,

$v = iR_0 + v_{OCR}$ (4.3)

This equation is represented in Fig. 4.7 by a line of slope $+R_0$ passing

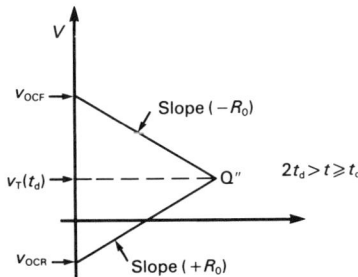

Fig. 4.7   Counterpart of Fig. 4.3 for reverse wave for $2t_d > t \geqslant t_d$.

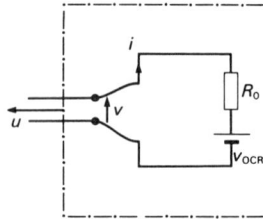

Fig. 4.8   SLL representation of reverse wave (compare with Fig. 4.4).

through the point Q'', having co-ordinates $v_T(t_d)$, $v_T(t_d)/R_T$. The plus sign for R_0 arises through the choice of sign discussed in Chapter 2: we wish to draw the characteristic of the reflected wave on the same diagram as we have used for the forward wave, in which the direction of current flow is taken as positive if flowing from left to right in the 'top' conductor.

To the left of the v_R wavefront the line appears as a battery of magnitude v_{OCF} in series with a resistor R_0. Fig. 4.7 should be compared with Fig. 4.3. Fig. 4.8 shows a sliding source equivalent circuit for v_R and should be compared with Fig. 4.4. From the geometry of Fig. 4.7 with $v_{OCF} = 2v_I(0+)$ $= 6$ V, and $v_I(t_d) = 2$ V, it follows that $v_{OCR} = -2$ V in the example considered. We can, if we wish, draw for v_R a diagram similar to that of Fig. 4.5 but this has not been done because no new principle is involved.

At the battery end of the line, Fig. 4.9(a) and Fig. 4.9(b) apply when the reflected wave arrives there. From either diagram $v_I(2t_d) = \{9 - (11 \times 100/150)\} = 5/3$ V, as predicted by the reflection chart of Fig. 3.5. The graphical construction can now be repeated for further forward and reverse waves. The intersections of successive lines of slope $+R_0$ with the source characteristic (in this case a battery) yield, respectively, $v_I(2t_d > t > 0+)$, $v_I(4t_d > t \geqslant 2t_d)$, etc., while the intersections of the lines with slope $-R_0$ with the load or receiving end characteristic give, respectively, $v_T(3t_d > t \geqslant t_d)$, $v_T(5t_d > t \geqslant 3t_d)$, etc. Theoretically, reflections occur until

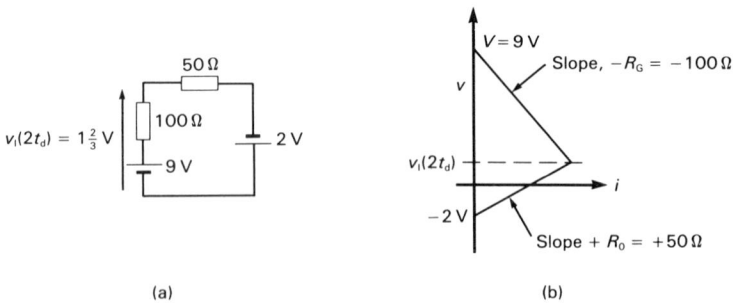

(a)   (b)

Fig. 4.9   Conditions at line input at $t = 2t_d$. (a) Equivalent circuit for finding $v_I(2t_d)$; (b) Graphical construction for (a).

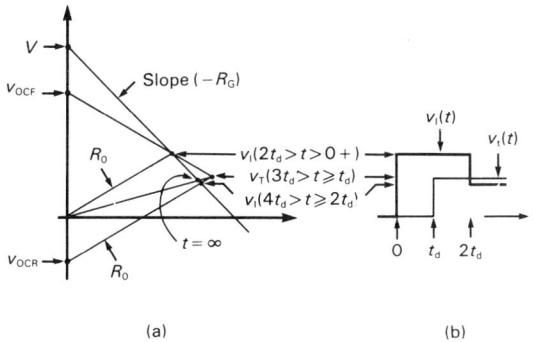

(a)   (b)

Fig. 4.10   Showing how waveforms at line input and load can be found.
(a) Composite time picture for $v_I(t)$, $v_T(t)$ for $4t_d > t > 0$ at $t = \infty$; (b) Waveforms deduced from (a).

$t = \infty$, at which time $v_I(\infty) = v_T(\infty) = 1.8$ V. This corresponds to the intersection of the output characteristic of the source with the load characteristic
at the end of the, supposedly lossless, cable.

To obtain $v_I(t)$ and $v_T(t)$ for $4t_d > t \geqslant 0+$, we can superpose the individual
time pictures to obtain, on a single v, i graph, the composite time picture
shown in Fig. 4.10(a). The required waveforms are readily derived by cross-
projection as indicated in Fig. 4.10(b). The lines with slope $\mp R_0$, used to
describe the forward and reverse waves are sometimes called *Bergeron lines* in
honour of the Frenchman who appears to have been the first to use this type
of graphical technique for describing the motion of (water) waves.

Two points should be remembered in using the Bergeron lines. Firstly, they
extend indefinitely in either direction in the v, i plane, i.e. they are not
confined to the half-plane $v > 0$. There is, though, little value in extending
them beyond the points at which they make intersections with the relevant

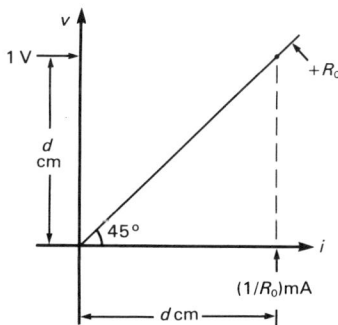

Fig. 4.11   A convenient choice of scale in plotting v, i characteristics
gives the $\pm R_0$ lines at right-angles.

source and load characteristics. Secondly, geometrical constructions using them are simplified, and problems involving them are solved rapidly using a set-square or protractor, if it can be arranged that they are at right angles. This is possible if the v, i plots for the source and load characteristics are scaled so that a conveniently chosen 'd' cm vertically corresponds to 1 V and 'd' cm horizontally corresponds to $(1/R_0)$ mA as shown in Fig. 4.11.

Example 4.1

Re-work Example 3.2 using an SLL approach.

Solution

The reader is referred to Fig. 3.6 for the circuit diagram. A graphical solution is given in Fig. 4.12. 'A' shows the source characteristic, 'B' the characteristic for the charged line, at $t = 0+$, and 'C' the load characteristic. This problem is relatively simple because the source resistance is equal to R_0.

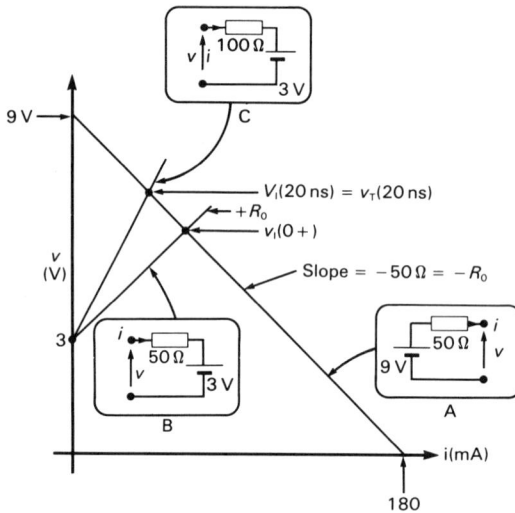

Fig. 4.12   Solution to Example 4.1 using SLL technique.

Example 4.2

In Fig. 4.13, Sw opens at $t = 0$. Using a SLL approach find $v_1(t)$ for 30 ns $> t > 0+$ and $v_T(t)$ for 20 ns $> t > 0+$.

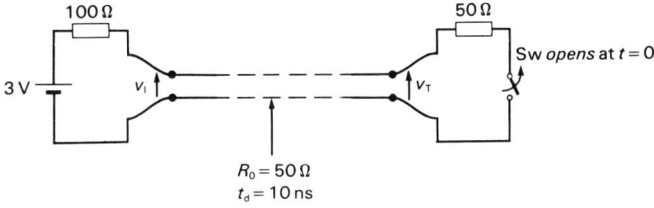

Fig. 4.13 Circuit for Example 4.2.

Solution

In Fig. 4.14(a), P gives the voltage across, and current in, the 50 Ω load resistor before Sw opens. The opening of Sw causes the current in the load to be zero and this corresponds to $v_T(0+) = 2$ V at point Q. The intersection of a line with slope $+R_0$, passing through Q, and the source characteristic gives $v_I(10$ ns). The required waveforms are shown in Fig. 4.14(b).

Though it offers some physical insight, the SLL technique would not normally be used with linear source and load resistance as the reflection

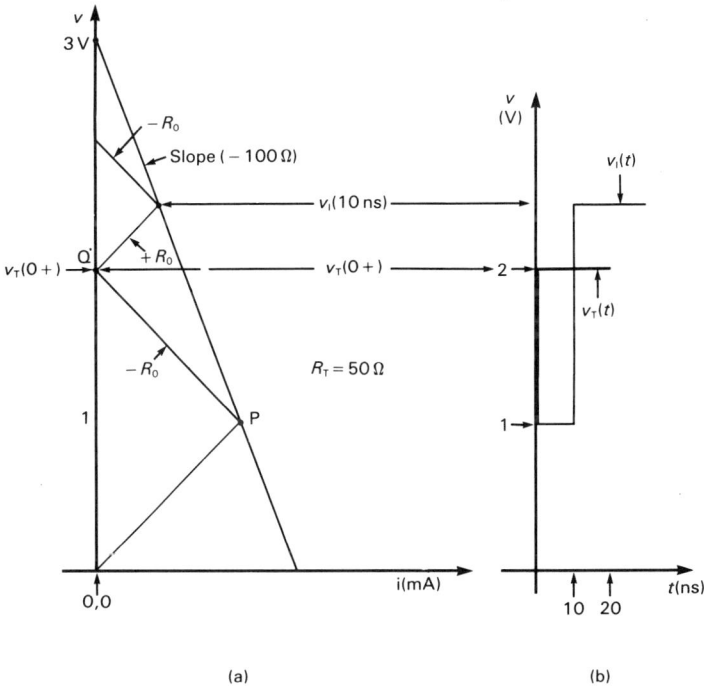

(a)   (b)

Fig. 4.14 Solution for Example 4.2.

chart is so quick and easy to apply. However, as emphasized earlier, the SLL technique comes into its own when dealing with nonlinear loads, an example of which is given in section 4.2. The sliding-source concept is also useful in interpreting results in time domain reflectometry (TDR), considered in Chapter 5.

4.2   NONLINEAR LOAD: A DIODE TERMINATION

Suppose we replace the resistor R_T in Fig. 4.1 by a semiconductor diode with its anode connected to the 'top' conductor and its cathode to the bottom one. The diode characteristic, as conventionally plotted, is shown in Fig. 4.15(a); for SLL analysis this is plotted again with v vertical and i horizontal as in Fig. 4.15(b).

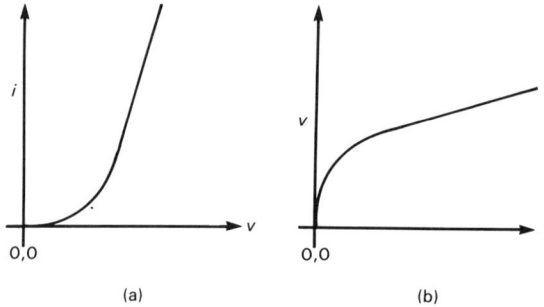

Fig. 4.15   Semiconductor diode characteristic. (a) Conventional representation; (b) plot used in SLL analysis.

None of the theory supporting the graphical construction of a sliding load line is altered — we do not alter the mechanism of wave propagation on the line, only the boundary conditions are different. $v_I(t)$ is given by the intersections of lines of slope $+R_0$ with the source characteristic $v = V - iR_G$ and $v_T(t)$ is obtained from the intersection of lines of slope $-R_0$ with the diode characteristic as drawn in Fig. 4.15(b).

Figure 4.16(a) is a composite time picture for $v_I(t)$, $v_T(t)$ for $2t_d > t \geqslant 0+$, for an arbitrary diode. Depending upon the conditions of the problem it may be possible to choose a diode which correctly terminates the line. This is easily seen graphically. The use of a diode to suppress overshoots on lines connecting digital hardware has been known for many years. As far as the author is aware no exact mathematical solution of the problem has yet been worked out.

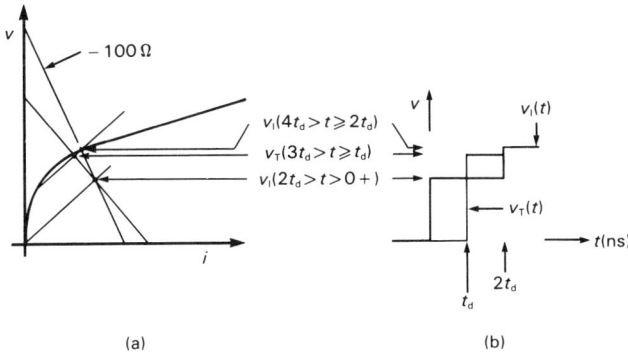

Fig. 4.16 (a) Shows composite time picture when semiconductor diode is load component; (b) Shows associated waveforms.

4.3 REFLECTIONS WITH LOGIC CIRCUIT INTERCONNECTIONS

For the preservation of pulse edges and the reduction of crosstalk effects (see Chapter 6) it is common practice to inter-connect logic-circuit elements with some form of transmission line. This usually means strip-line on a printed circuit board and twisted-pair or coaxial cables between board assemblies.

Thus the problem reduces to investigating reflection effects which occur and seeing if we can 'live' with them. This will be accomplished if any reflections cause no voltage overstress in the driving and driven gates and no logic malfunction, through spurious triggering, of other gates to which they are connected.

We consider first transistor–transistor logic (TTL) then emitter-coupled logic (ECL). The d.c. characteristics of a TTL gate are shown in Fig. 4.17. The positive logic convention is assumed, i.e. the more positive of the two discrete voltage levels in the system is taken as '1'.

The curves may be obtained from manufacturers' data sheets or generated experimentally using an x, y plotter. A single continuous curve, (i) in Fig. 4.17(a), gives $v_1 = f(i_1)$. For each value of v_1 there is a single value of v_0, but we are interested in only the v, i characteristics corresponding to $v_0 \equiv$ '0' and $v_0 \equiv$ '1'. These are shown, respectively, as curves (ii) and (iii) in Fig. 4.17(b), (c).

Bearing in mind our v, i sign convention we now superpose (i), (ii), (iii) to obtain the TTL composite d.c. characteristics (Fig. 4.19) relevant to the two cascaded gates, A and B in Fig. 4.18, connected via a length of transmission line with a one-way delay time t_d.

We consider only the '1' → '0' transition at the line input — the case of the '0' → '1' transition follows by analogous treatment. Thus, the output of A is initially '1' and has been so long enough for any past line transients, which

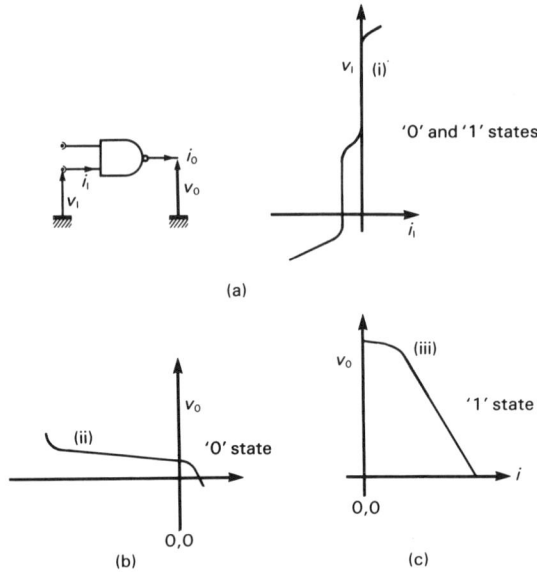

Fig. 4.17 (a) TTL gate showing sign convention and input characteristic; (b) '0' level output characteristic; (c) '1' level output characteristic.

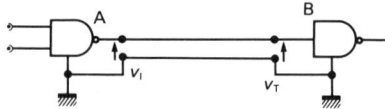

Fig. 4.18 Interconnected TTL gates.

may have occurred, to have died away; then $v_I = v_T = v_a$ and this corresponds to 'a', the point of intersection of (i) and (iii). The dynamic representation of the line at $t = 0$, when v_I switches to the '0' condition, is simply a resistor R_0 in series with a battery v_a corresponding to the equivalent circuit of a charged line. This is given by the straight line of slope $+R_0$ passing through 'a'.

Also, $v_I(0+) = v_I(2t_d \geqslant t \geqslant 0+) = v_b$, where 'b' is the intersection of this load line, extended back, with (ii). The sliding-source circuit for the wavefront which starts down the line is a battery v_{OCF} in series with a resistor R_0. This is depicted graphically by a line with slope $-R_0$ passing through 'b'. Where this intersects (i) gives $v_T(t_d) = v_T(3t_d > t \geqslant t_d) = v_c$. We can continue to draw our Bergeron lines and thus determine $v_I(t)$, $v_T(t)$ for whatever time range we choose. We have enough information, of course, to plot $v(t)$ for any point $x = x'$ on the line but this is rarely required. Plots of $v_I(t)$, $v_T(t)$ for $2t_d \geqslant t \geqslant 0+$ are derived simply and are given in Fig. 4.20.

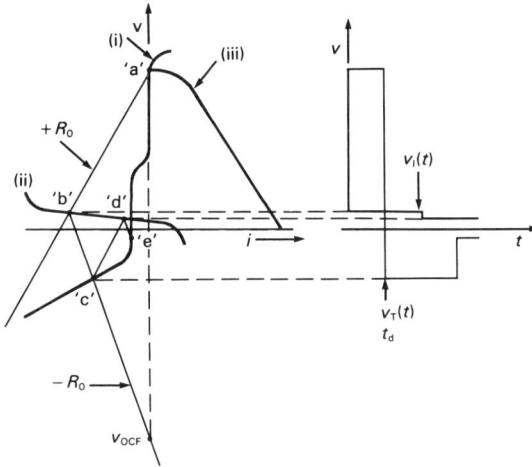

Fig. 4.19 Composite d.c.
characteristics of TTL gate with
'Bergeron' lines.

Fig. 4.20 Waveforms for
$2t_d \geqslant t \geqslant 0+$ deduced from
Fig. 4.19.

The first negative-going excursion in $v_I(t)$ can, if large enough, lead to a subsequent overshoot in v_T sufficient to switch B. Various methods of reducing the likelihood of this happening have been described. They include the careful control of the input characteristic in the design of the TTL gate, often by the use of some form of built-in clamping diode. Provided line pulse reflections cause no gate malfunction and have effectively 'died-out' before the next driving-gate output transition occurs, the absence of precise line matching with TTL is unimportant.

In practical digital system design, TTL is useful for clock frequencies of some tens of megahertz. Operation at the highest clock frequencies requires a logic gate having sub-nanosecond switching times and input and output characteristics that permit the use of matched lines. This means the use of an emitter-coupled logic (ECL) gate, a symbol for which is shown in Fig. 4.21(a). The incremental input resistance of an ECL gate, over its useful input range, exceeds several kΩ, i.e. very much greater than the R_0 of a line connected to it. Furthermore, the direct input current is only a few μA. Thus, (i) on the vertical axis in Fig. 4.21(b) is a good approximation to the d.c. input characteristic.

The emitter-follower outputs of an ECL gate are not always supplied with built-in load resistors; we will assume they are absent in the discussion that follows. However, a d.c. return path to the -5.2 V rail or an auxiliary rail is required and must be made via externally connected components. Then the effective output characteristics in the '1' and '0' logic states are curves (ii) and (iii), respectively. These, which apply for both the OR and NOR outputs, are

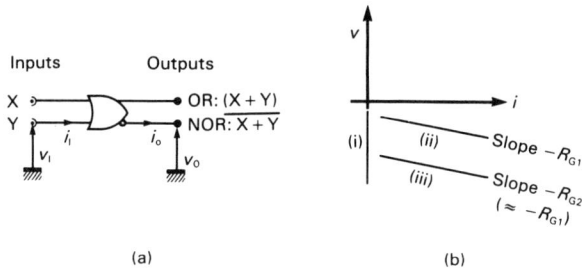

(a)   (b)

Fig. 4.21 ECL gate. (a) Symbol; (b) Gate characteristics: (i) Input characteristic, (ii) Output characteristic '1' state, (iii) Output characteristic '0' state.

almost parallel to the i axis. The mean output resistances R_{G1}, R_{G2} account for the finite slope. Typically, $R_{G1} \approx R_{G2} \approx 10~\Omega$: average values for the '1' and '0' levels are -0.8 V, -1.75 V respectively when 0 and -5.2 V rails are used. Either receiving-end (parallel) matching or sending-end (series) matching can be employed.

Figure 4.22(a) shows parallel matching with a fan-out of 3. Connecting $R_T(=R_0)$ to a -2 V rail avoids a large standing current in the driving gate when it is at the '0' level. If an auxiliary -2 V rail is not available the Thévenin equivalent termination shown in Fig. 4.22(b) can be employed.

With parallel matching the full logic swing is propagated along the line. This can be seen by reference to Fig. 4.23(a). A straight line with slope $R_T = +R_0$, passing through the point $v = -2$ V, $i = 0$, describes the load characteristic. It cuts curve (iii) at 'a' so v_a gives the line voltage in the '0' state. When a '0' → '1' transition occurs at the output of the gate the relevant output characteristic becomes (ii). This cuts the charged line characteristic at v_b corresponding to the sending end voltage in the '1' state. A straight line of slope $-R_0$, passing through 'b' describes the motion of the '0' → '1'

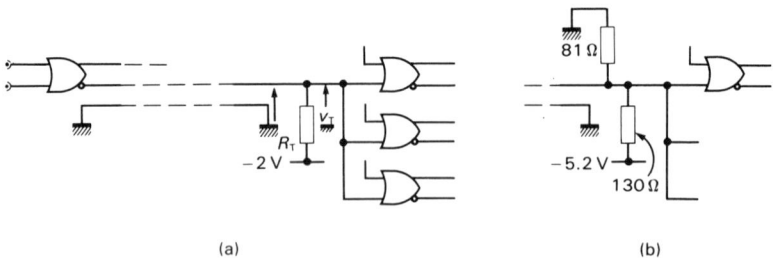

(a)   (b)

Fig. 4.22 Parallel termination for ECL. (a) Using an auxilliary supply; (b) Using a potential divides network: the two resistors are equivalent to 50 Ω connected to -2 V.

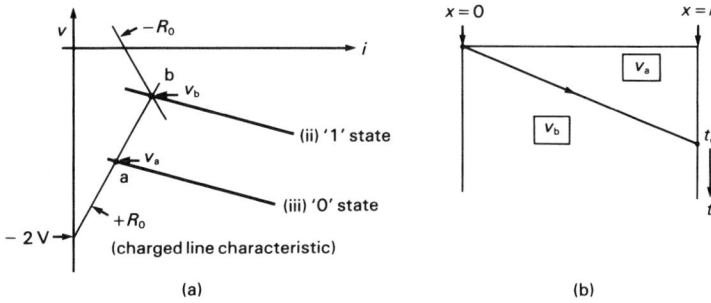

Fig. 4.23   Characterization   of   '0'→'1'   transition.   (a) Using   SLL
approach; (b) Using Reflection Chart.

transition down the line. This Bergeron line intersects the load characteristic
at 'b' and there is no reflection.

The reflection chart description of the phenomenon is, of course, much
simpler and is shown in Fig. 4.23(b). Note that for parallel termination the
terminating resistor provides a satisfactory load for the emitter-follower of
the driving gate so it is not obligatory to include a load resistor at the sending
end. An SLL analysis of the '1' → '0' transition is analogous to that for the
'0' → '1' transition.

It is possible to 'fan-out' not only at the end, but also at various points
along a parallel-terminated line (see Fig. 4.24). If the line is loaded at
approximately equal intervals and the stubs to the driven gates are kept short,
then it is acceptable, as a first-order approximation, to regard the lumped
input capacitance of each gate as uniformly distributed along the line.

For n gates each having an average input capacitance C_G the equivalent
distributed loading is C_L per unit length.

$$C_L = (nC_G/l) \tag{4.4}$$

The modified line parameters are:

$$R_0' = R_0/\sqrt{\{1 + (C_L/C)\}} \tag{4.5}$$

$$t_u' = t_u\sqrt{\{1 + (C_L/C)\}} \tag{4.6}$$

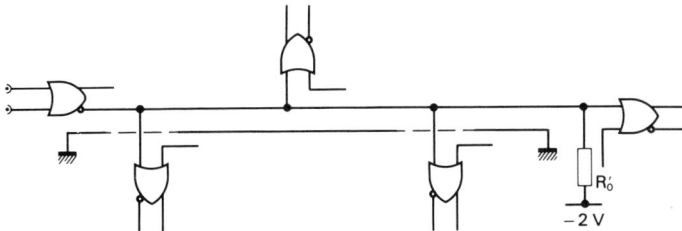

Fig. 4.24   Line with distributed gate loading.

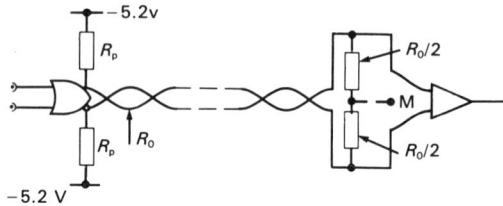

Fig. 4.25   ECL gate connected to a line receiver.

where C is, of course, the unloaded line capacitance per unit length.

Parallel termination is useful when a gate is connected to an ECL line receiver via a twisted-pair transmission line, as shown in Fig. 4.25. If the point M is connected to -2 V the resistors R_p can be omitted. However, if M is left unconnected, or is connected to earth potential, the resistors R_p are necessary.

Figure 4.26 shows series matching. Only one-half the logic swing is propagated along the line and is doubled due to reflection at the open end. The full logic swing is achieved at the driving end, after a time $2t_d$, and there are no reflections from it if

$$[R_S + \{R_p R_G/(R_p + R_G)\}] = R_0 \tag{4.7}$$

Normally $R_p \gg R_G$ so this reduces to

$$(R_S + R_G) = R_0 \tag{4.8}$$

With series matching, loads cannot be distributed along a line. However, several lines each with a series resistor at the sending end can be driven from the output of an ECL gate. A limit is set by the power dissipation in the ECL output stage.

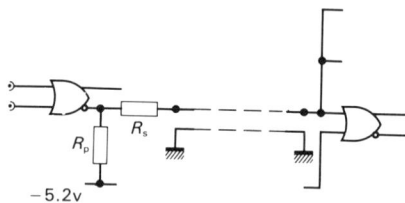

Fig. 4.26   ECL gate with series matching.

PROBLEMS

Use the SLL approach in solving these problems unless otherwise indicated.

P4.1   In Fig. P4.1, Sw closes at $t = 0$. Sketch $v_1(t)$ and $v_T(t)$ for

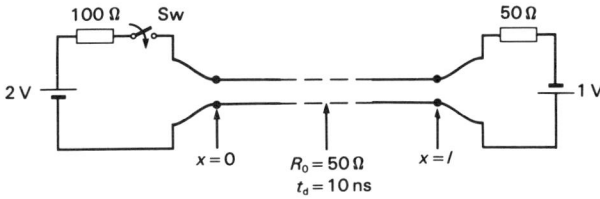

Fig. P4.1.

30 ns $> t > 0+$. Check your results using a reflection chart.

P4.2 In Fig. P4.2:

a Sw is initially open and closes at $t = 0$. Sketch $v_X(t)$ for 40 ns $> t > 0+$.

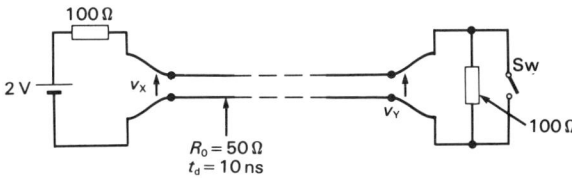

Fig. P4.2.

b Sw is initially closed and opens at $t = 0$. Sketch $v_X(t)$, $v_Y(t)$, for 25 ns $\geqslant t > 0+$.

P4.3 In Fig. P4.3:

a Sw is initially closed and opens at $t = 0$. Show that $v_Y(0+) = 1.33$ V.

b Sw is initially open and closes at $t = 0$. Sketch $v_X(t)$, $v_Y(t)$ for 30 ns $> t > 0+$.

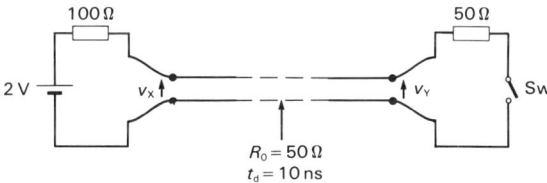

Fig. P4.3.

P4.4 Fig. P4.4 shows, schematically, a fast pulse generator in which Sw represents a solenoid-actuated, mercury-wetted relay switch capable of making extremely fast contact without the 'bounce' normally encountered with traditional mechanical switches. Sw is initially open and the co-axial cable type RG8A/U, for which $t_u \approx 1.475$ ns/ft, is charged up to the rail voltage $+100$ V. Sketch $v_Z(t)$

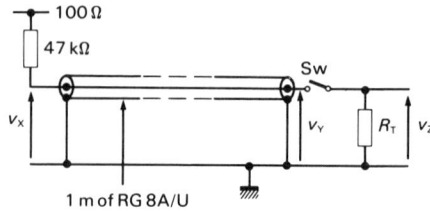

Fig. P4.4.

for $12 \text{ ns} > t > 0+$ for $R_T = 50 \ \Omega$ if S_w closes at $t = 0$. (Use any approach in your solution.)

P4.5   Repeat P4.4 for $R_T = 25 \ \Omega$, all other data remaining the same.

P4.6   Referring to Fig. P4.4, sketch the waveform $v_X(t)$ when Sw, assumed initially closed, subsequently opens. (Hint: use with a justifying argument, the formula presented in P3.10(b)). Assume $C = 29.5 \text{ pF/ft}$ for RG8A/U. What limitations does this waveform place on the pulse repetition frequency of the pulse developed across R_T when Sw closes and opens repeatedly?

P4.7   Referring to Fig. 4.18 and the composite TTL characteristics of Fig. 4.19, sketch $v_I(t)$, $v_T(t)$ for $2t_d \geqslant t > 0+$, when the output of gate A makes a '0' → '1' transition.

P4.8   Show with the aid of sketches how $v_T(t)$ can be found for the ECL gate set-up of Fig. 4.22(a), when $R_T = 25 \ \Omega$ and, the driving gate output makes: a '0' → '1' transition; a '1' → '0' transition. (Assume $R_0 = 50 \ \Omega$.)

P4.9   Show that the ECL 'branching circuit' of Fig. P4.9 is correctly matched. (Note: for ease of drawing, only one wire of each transmission line is shown.)

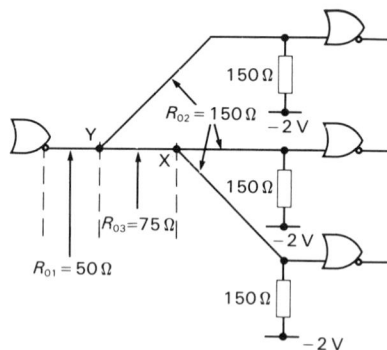

Fig. P4.9.

P4.10   Explain ECL series matching, using the SLL approach.

5 Time domain reflectometry

The abbreviation TDR stands for the instrument (Time Domain Reflecto-meter) and the technique (Time Domain Reflectometry). The context in which the letters are used generally make it obvious which is meant.

A time domain reflectometer is based on a simple closed-loop pulse radar, or echo-sounding system. It is used to test a system involving transmission lines and their accessories. It can be employed to determine: cable charac-teristic impedance; the nature and magnitude of components comprising a terminating impedance; the location and nature of faults or discontinuities; the properties of cable connectors and related components.

A schematic set-up of a TDR is shown in Fig. 5.1. A step-voltage wave-form is sent down a line and the incident and reflected waveforms are monitored by probes and displayed on the screen of a sampling oscilloscope having an effective bandwidth that will normally need to be measured in gigahertz (GHz), corresponding to a rise-time of a few tens of picoseconds.

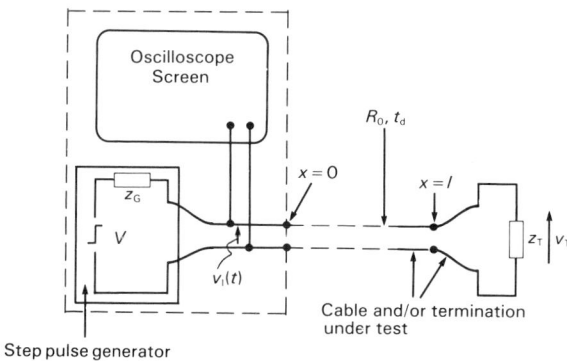

Fig. 5.1 Basic TDR scheme. Normally $Z_G = R_0 = 50\ \Omega$.

The pictorial display makes it easy to interpret the results if, as is often the case, the output resistance of the step generator matches the nominal characteristic impedance of the line under test and the rise-time of the 'step' waveform is less than 50 ps.

The interpretation of a TDR display is best understood by the examples below. These also illustrate the application of the concepts of preceding Chapters. The problems amount to finding $v_I(t)$, i.e. the time variation of line voltage at the line input, so details not specifically related to that are omitted.

By virtue of the function it performs, the TDR is usually an expensive instrument. Nevertheless, the technique of TDR can be demonstrated in practice using a standard laboratory oscilloscope (bandwidth > 30 MHz) and pulse generator, if transmission lines of sufficient length are used (see Appendix B).

5.1 APPLICATION EXAMPLES: STEP AND TRUNCATED-RAMP DRIVE

Example 5.1

In Fig. 5.2 a line of characteristic impedance R_0 and one-way delay time t_d is terminated by a discrete capacitor, C_L, assumed ideal. Determine $v_I(t)$, as would be observed by a TDR for which the step generator v_G has an output impedance $Z_G = R_0$.

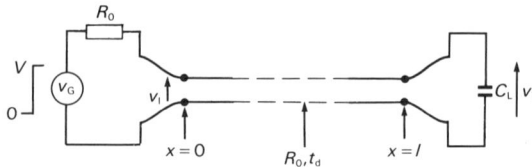

Fig. 5.2 v_G is the step-generator of a TDR; $v_I(t)$ is displayed on oscilloscope screen.

Solution

The diagrams of Fig. 5.3 illustrate line behaviour. The initial forward wave $v_f = v_I(0+)$ that travels down the line has an amplitude $(V/2)$: refer to (a). Comparing this with Fig. 2.16, $f(t) = (V/2) U(t)$.

The sliding equivalent source representation of conditions at the end of the line when v_f reaches it is given in (b).

The equation for $v_T(t)$ is,

$$U(t - t_d)V = iR_0 + v_T \tag{5.1}$$

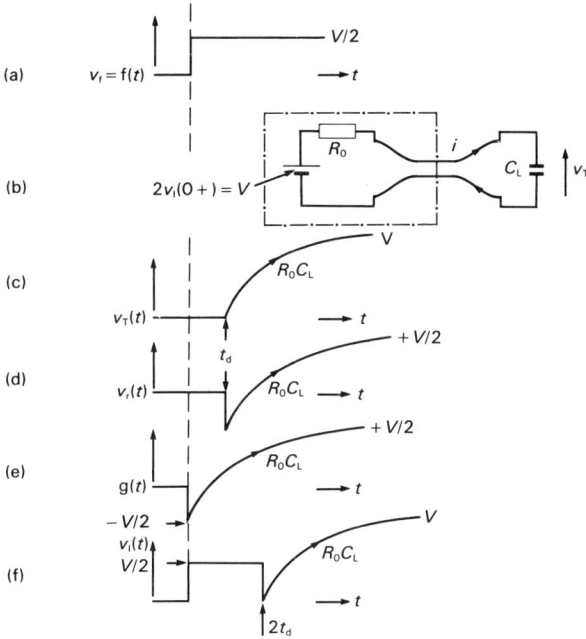

Fig. 5.3 Relating to Example 5.1. (a) Input step $= f(t)$; (b) Dotted box gives equivalent source representation for line at $t = t_d$; (c) $v_T(t)$; (d) $v_r(t)$; (e) $g(t)$; (f) $v_I(t)$.

The function $U(t - t_d)$ allows for line delay (see Chapter 2).

$$U(t - t_d) = 1, t > t_d$$
$$U(t - t_d) = 0, t < t_d$$

We can also write,

$$i = C_L(dv_T/dt) \qquad (5.2)$$

From eqns (5.1) and (5.2),

$$U(t - t_d)V = v_T + R_0 C_L (dv_T/dt) \qquad (5.3)$$

The solution of eqn (5.3) is a delayed exponential waveform having time-constant $R_0 C_L$ and amplitude V, as shown in (c).

Thus, $v_T(t - t_d) = U(t - t_d)[V\{1 - \exp - (t - t_d)/R_0 C_L\}] \qquad (5.4)$

Now, $v_T(t - t_d) = v_f(t - t_d) + v_r(t - t_d) \qquad (5.5)$

$\therefore \quad v_r(t - t_d) = U(t - t_d)[V\{1 - \exp - (t - t_d)/R_0 C_L\} - V/2] \qquad (5.6)$

This is shown in (d).

If we refer again to Fig. 2.16, it is as if the waveform g(t) for the phantom voltage generator is that given in (e).

$$g(t) = U(t)[V\{1 - \exp(-t/R_0 C_L)\} - V/2] \tag{5.7}$$

The waveform for $v_1(t)$, (see (f)) is the sum of f(t) and $U(t - 2t_d)g(t - 2t_d)$.
In mathematical form,

$$v_1(t) = (V/2) U(t) + U(t - 2t_d)[V\{1 - \exp - (t - 2t_d)/R_0 C_L\} - V/2] \tag{5.8}$$

$v_1(2t_d) = 0$ because the terminating capacitor, C_L, behaves initially as a short-circuit. $v_1(t \to \infty) = V$ because, under d.c. conditions, C_L appears as an open-circuit; Fig. 5.3(f) is easily catalogued as the 'signature' for a capacitive termination. Hence t_d follows by inspection because the oscilloscope screen is scaled horizontally in ps/cm or ns/cm, etc. and C_L can be estimated from the time taken by the exponential section of the waveform to reach $V/2$, if R_0 is known. (See problem P5.1.)

An expression for $v_1(t)$ can also be obtained using a Reflection Chart approach as the system is linear. We can find an expression for $\rho_{VT}(s)$, s being the Laplace Transform variable, and given the step-function nature of v_f find v_r and hence v_1. (See problem P5.2.) However, the solution offered here does give more physical insight.

Example 5.2

Figure 5.4 shows a line correctly terminated at its receiving end but having a small shunt capacitive loading at a point $x = x'$ along its length. Determine $v_1(t)$, as might be observed using a TDR.

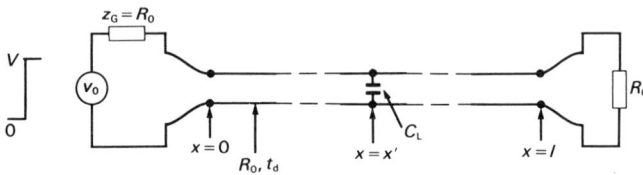

Fig. 5.4   Capacitive loading at a point on a line correctly terminated at the receiving end.

Solution

The approach is similar to that of Example 5.1. Now, however, a reflection occurs at the discontinuity at $x = x'$. The diagrams of Fig. 5.5 apply.

The Thévenin equivalent circuit that applies when the forward wave meets the discontinuity is shown in (b). R_0, shunting C_L, refers to the incremental

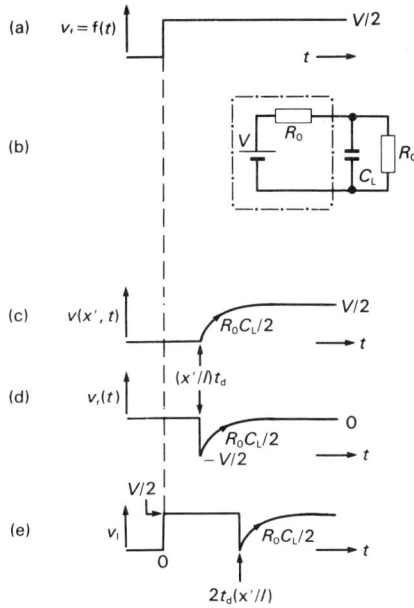

Fig. 5.5 Relating to Example 5.2.

impedance looking to the right from C_L. The time-constant associated with the waveform at the discontinuity is now $R_0 C_L/2$ as shown in (c). The waveform, v_r, reflected back towards $x = 0$ from the discontinuity is the difference between v_f and $v(x', t)$, see (d).

Now $v_I(t)$, as indicated in (e), is the sum of (a) and (d) delayed. Step input edges do not always apply in practice. Consider Fig. 5.6(a) in which an impedance Z is connected in parallel with the line at x'. $v_G(t)$ is the truncated-ramp input waveform shown in (b). Suppose, first that Z is a pure resistor. To simplify the arithmetic, let $Z = R_0$. The relevant diagrams appropriate to the problem of finding $v_I(t)$ are shown in Fig. 5.7. Figs 5.7(a)–(e) should be compared with Figs 5.5(a)–(f).

Suppose, next, that Z is a pure capacitor, C_L. Relevant diagrams are given

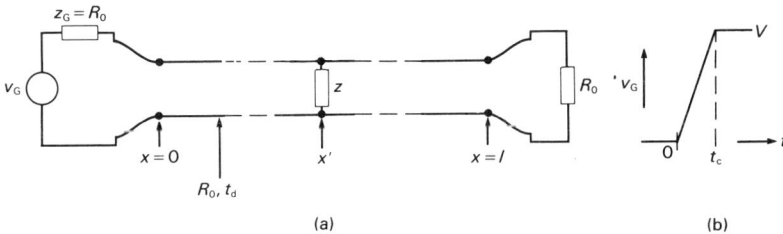

Fig. 5.6 (a) Line with impedance Z at point x'; (b) Generator waveform for (a).

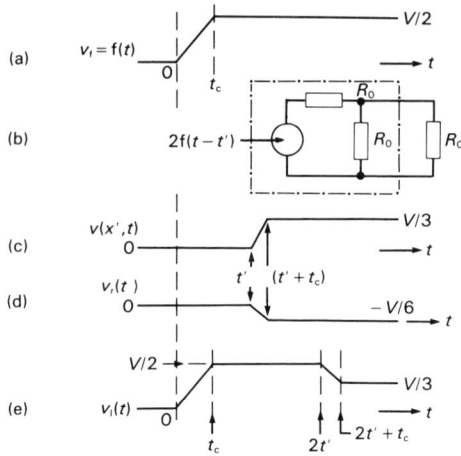

Fig. 5.7   Waveforms and equivalent circuit for Fig. 5.6 for $Z = R_0$;
$t' = (t_d x'/l)$.

in Fig. 5.8: (a) is the incident waveform v_f; $v(x', t)$, calculated from the equivalent circuit of (b) is shown in (c); v_r, the difference between (c) and (a) is depicted in (d). For $t > t'$ ($= t_d x'/l$), $v(x, t) > 0$ because C_L does not appear as a short-circuit for a waveform with a finite rise-time.

Assuming $t_c \gg (R_0 C_L/2)$, it can be shown (problem P5.6) that

$$|v_r|_{max} \approx (V/4t_c)(R_0 C_L) \qquad\qquad\qquad (5.9)$$

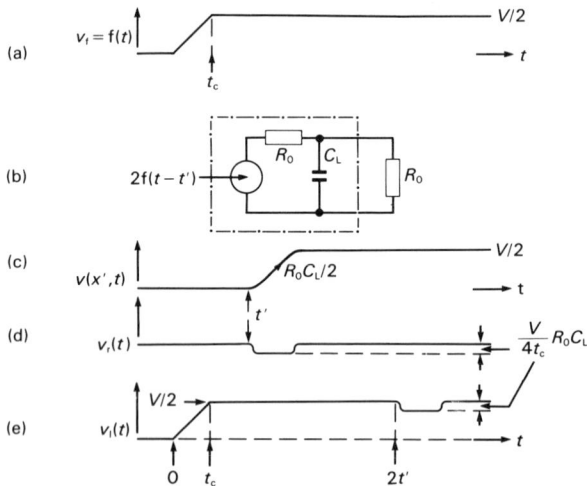

Fig. 5.8   Waveforms and equivalent circuit for Fig. 5.6 for $Z = C_L$;
$t' = (t_d x'/l)$.

The shape of $v_1(t)$ provides a means of locating the position along the line, and estimating the magnitude, of C_L.

PROBLEMS

Where reference is made to Fig. 5.1, assume that $Z_G = R_0$.

P5.1 For the set-up of Fig. 5.2, $C_L = 100$ pF and $R_0 = 50\ \Omega$. Calculate $t_k (= t - 2t_d)$ for which $v_1(t_k) = V/2$.

P5.2 The input waveform to a line $(x = 0)$ is $v(0, t) = v_1(t) = f(t)$. Show that a description of the forward waveform $v_f(x', t)$ in the notation of the Laplace transform is given by:
$v_f(x', s) = f(s) \exp(-sx't_u) = f(s)\exp(-st')$, where $t' = (x'/u)$. (Hint: Refer to eqn (2.11) of Chapter 2.)

P5.3 Derive an expression for $v_1(s)$, and hence $v_1(t)$, for Example 5.1 of the text using a reflection chart approach and the s-domain description of ρ_{VT}, i.e. $\rho_{VT}(s) = \{Z_T(s) - R_0\}/\{Z_T(s) + R_0\}$.

P5.4 Sketch waveforms similar to those of Fig. 5.3(a) to (f) for each of the terminations A, B, C shown in Fig. P5.4.

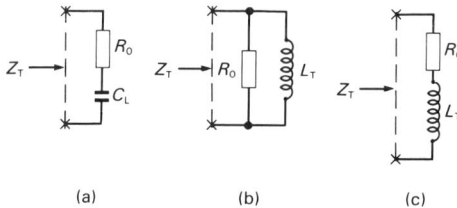

(a)   (b)   (c)

Fig. P5.4 Terminating impedances for problem P5.4.

P5.5 Using the sketches of Fig. P5.4, write down mathematical expressions for $v_1(t)$, $v_T(t)$ for each of the terminations indicated.

P5.6 Derive eqn (5.9) of the text.

6 Crosstalk

Crosstalk is the name given to unwanted signals generated in a line by transient signals in an adjacent line. The physical proximity of the lines results in electrical coupling, both inductive and capacitive. Crosstalk can occur with strip lines on printed circuit boards (p.c.b.s) and with cable runs involving bundles of lines. We will consider each in turn. However the main emphasis is on strip lines, which are more amenable to mathematical treatment.

6.1 CROSSTALK WITH STRIP LINES

This is an important practical case and lends itself to quantitative analysis. In Fig. 6.1 two strip lines formed on a printed circuit board, share a common ground plane. Line B, the 'pick-up' or 'sensor' line has a top copper track that runs parallel for a distance l, to the top track of line A, the 'driver' line

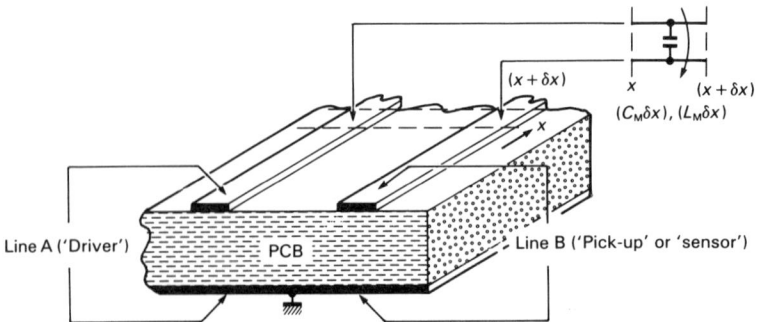

Fig. 6.1   Coupling with strip lines.

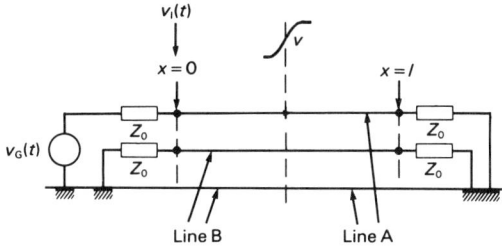

Fig. 6.2   Schematic representation of Fig. 6.1 with drive source and
both lines matched at each end.

along which a pulse is propagated. Because the two top conductors are close
to each other there is a mutual capacitance, C_M, and a mutual inductance, L_M,
per-unit-length associated with the two lines. Thus, for a coupled-line section
between x and $(x + \delta x)$ there is a mutual capacitance $(C_M \delta x)$ and a mutual
inductance $(L_M \delta x)$.

Figure 6.2 shows, in schematic form, the two lines A and B each matched at
each end by its characteristic impedance Z_0. A rigorous mathematical
analysis of even this special case is quite tedious, but is simplified if it is
assumed that only 'weak' coupling exists between the lines. This means,
effectively, that *line B is affected by but does not, in turn, affect line A* when a
drive signal exists in the latter. In practice this requires that $C_M \ll C$, $L_M \ll L$
and this implies that each line has the characteristic impedance it would have
if the other line were absent.

These assumptions are justified to the extent that observed and predicted
results are in reasonable agreement in many practically interesting cases.

Consider Fig. 6.3(a) and the incremental components of voltage induced
into the sensor line between x and $(x + \delta x)$ owing to capacitive coupling.

The induced current is δi_c, where

$$\delta i_c = (C_M \delta x)(\delta v / \delta t) \qquad (6.1)$$

Now δi_c 'sees' an incremental impedance Z_0 looking in each direction along

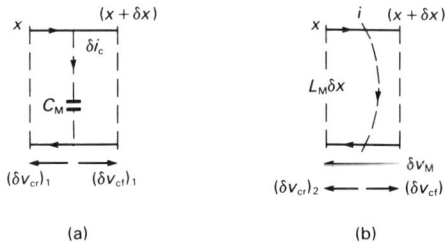

(a)   (b)

Fig. 6.3   Voltage contributions to forward crosstalk component δv_{cf} and
reverse crosstalk component δv_{cr} resulting from: (a) Capacitive coupling;
(b) Inductive coupling.

line B and divides equally to produce oppositely directed components $(\delta v_{cf})_1$, $(\delta v_{cr})_1$), of forward and reverse crosstalk voltage, respectively.

$$(\delta v_{cf})_1 = (\delta v_{cr})_1 = (i_c Z_0/2) = (C_M \delta x)(\delta v/\delta t)Z_0/2 \tag{6.2}$$

Consider, next, the incremental components of voltage induced into the sensor line between x and $(x + \delta x)$ owing to inductive coupling. The induced voltage, δv_m, has a directional property shown by the arrow in Fig. 6.3(b). This is because Lenz's Law requires it to act in a direction opposing the change of flux between the two lines.

$$\delta v_m = (L_M \delta x)(\delta i/\delta t) \tag{6.3}$$

or, as $i = v/Z_0$,

$$\delta v_m = (L_M \delta x)(\delta v/\delta t)/Z_0 \tag{6.4}$$

δv_m can be resolved into two opposite travelling components, $(\delta v_{cf})_2$, $(\delta v_{cr})_2$.

$$(\delta v_{cf})_2 = -(\delta v_{cr})_2 = -(L_M \delta x)(\delta v/\delta t)/2Z_0 \tag{6.5}$$

Thus, for the section of coupled line between x and $(x + \delta x)$, we have resultant forward and reverse components (δv_{cf}), (δv_{cr})

$$(\delta v_{cf}) = (\delta v_{cf})_1 + (\delta v_{cf})_2 \tag{6.6(a)}$$

$$(\delta v_{cr}) = (\delta v_{cr})_1 + (\delta v_{cr})_2 \tag{6.6(b)}$$

From eqns (6.2) and (6.5),

$$(\delta v_{cf}) = \{(C_M Z_0/2) - (L_M/2Z_0)\} \delta x(\delta v/\delta t) \tag{6.7(a)}$$

$$(\delta v_{cr}) = \{(C_M Z_0/2) + (L_M/2Z_0)\} \delta x(\delta v/\delta t) \tag{6.7(b)}$$

Figure 6.4 shows the crosstalk contributions resulting from a typical drive waveform having an 'S-shaped' wavefront. δv_{cr}, δv_{cf} have the same shape since they are both related to the time derivative of v. However, in practice, it is normally true that $(L_M/Z_0) > (C_M Z_0)$ so the polarity of δv_{cf} is opposite that of δv_{cr}.

We now consider the crosstalk waveforms in more detail.

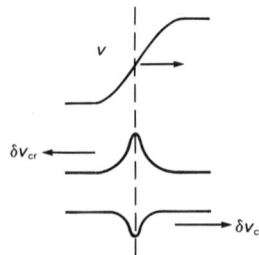

Fig. 6.4   Crosstalk contributions for a typical drive waveform.

6.1.1 Forward crosstalk waveform: derivation

From eqn (6.7(a)),

$$v_{cf}(x, t) = \Sigma \delta v_{cf} = \Sigma K_{cf} \left(\frac{\delta v}{\delta t} \right) \delta x \qquad (6.8)$$

where K_{cf} = forward crosstalk coupling coefficient. K_{cf} has the dimensions (T/L): convenient units are ns/ft.

$$K_{cf} \triangleq \{(C_M Z_0/2) - (L_M/2Z_0)\} \qquad (6.9)$$

In the limit case $\delta x \rightarrow 0$, $\delta t \rightarrow 0$, eqn (6.8) becomes

$$V_{cf}(x, t) = K_{cf} \int \left(\frac{dv}{dt} \right) dx \qquad (6.10)$$

Now, v results from the propagation at constant velocity of the driving signal v_I down line A.

$$\text{Hence, } v = U \left(t - \frac{x}{u} \right) v_I \left(t - \frac{x}{u} \right) \qquad (6.11(a))$$

To minimize algebraic complexity we will take the Heaviside shifting operator $U \left(t - \dfrac{x}{u} \right)$ as understood and write,

$$v = v_I \left(t - \frac{x}{u} \right) \qquad (6.11(b))$$

The shape of v remains constant as the drive signal progresses, because we are assuming lossless lines, so (dv/dt) for any point on the waveform remains constant.

Hence (dv/dt) can be taken outside the integral sign in eqn (6.10) to give,

$$v_{cf}(x, t) = K_{cf} \frac{d}{dt} \left\{ v_I \left(t - \frac{x}{u} \right) \right\} \int_0^x dx \qquad (6.12)$$

$$\text{or, } v_{cf}(x, t) = K_{cf} x \frac{d}{dt} \left\{ v_I \left(t - \frac{x}{u} \right) \right\} \qquad (6.13)$$

Physically, v_{cf} maintains its shape but grows in amplitude in proportion to the distance covered because the induced voltage components add in time synchronism. Usually the induced voltage at $x = l$ (corresponding to $t = t_d$) is of most interest.

$$v_{cf}(0, t) = K_{cf} l \frac{d}{dt} \{v_I(t - t_d)\} \qquad (6.14)$$

A mechanical analogy – the bulk transport of solid particles – helps in the understanding of the generation, and progress, of v_{cf}. In Fig. 6.5 sand falls

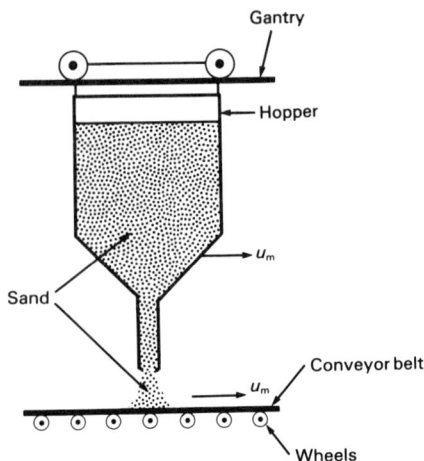

Fig. 6.5 Mechanical analogy for generation of forward travelling cross-talk. u_m = speed of hopper and conveyor belt.

vertically from a hopper, supported by an overhead gantry, on to a conveyor belt. If the gantry causes the hopper to move in the same direction and at the same speed over the belt a small pile of sand will build up at only one part of the conveyor belt, i.e. directly beneath the hopper exit pipe, and its size will depend on the distance travelled.

In this analogy the sand/hopper arrangement is playing the part of the drive waveform on line A and the conveyor belt with its pile of deposited sand plays the part of the forward crosstalk waveform on line B.

6.1.2 Reverse crosstalk waveform: derivation

The reverse crosstalk waveform at a point x' on the sensor line is the time integral of all the increments of induced voltage received in time-sequence at that point after the drive signal, on line A, has passed x'. The increments continue until the drive signal has reached $x = l$. Though no reflections are involved with the set-up of Fig. 6.2, the line conditions associated with the generation of a reverse crosstalk waveform are more easily understood using a reflection chart.

A reverse crosstalk component δv_{cr1} is generated at x' on line B as the drive signal on line A passes the point (Fig. 6.6). But, during the time interval δt in which v moves from x' to $(x' + \delta x')$, and in so doing generates a reverse cross-talk component δv_{cr2}, δv_{cr1} has moved a distance $\delta x'$ in the opposite direction. The net contribution to the reverse crosstalk waveform at x' is thus $(\delta v_{cr}/2)$.

Using eqn (6.7(b)) we can write,

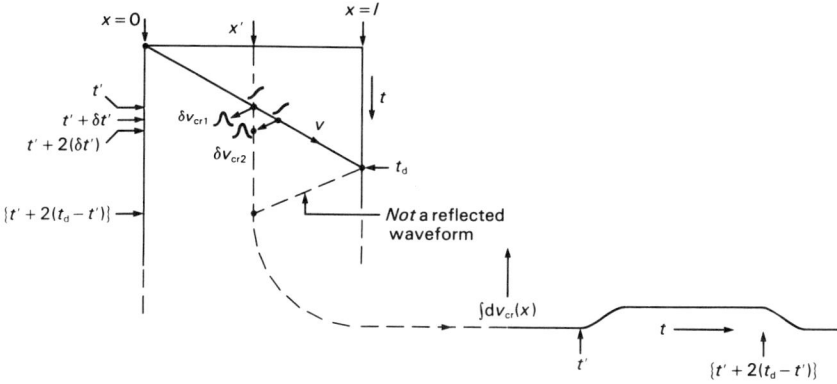

Fig. 6.6 Generation of reverse crosstalk waveform by drive signal v.

$$v_{cr}(x, t) = \lambda \Sigma \left(\frac{\delta v}{\delta t} \right) \delta x \tag{6.15}$$

where, $\lambda \triangleq 1/4 \{(C_M Z_0) + (L_M/Z_0)\}$ (6.16)

In the limit, eqn (6.15) becomes:

$$v_{cr}(x, t) = \lambda \int \left(\frac{dv}{dt} \right) \left(\frac{dx}{dt} \right) dt \tag{6.17}$$

Now for the reverse crosstalk waveform,

$$\left(\frac{dx}{dt} \right) = -u \tag{6.18}$$

$$\text{Hence, } v_{cr}(x, t) = -\lambda u \int_{v_1}^{v_2} dv \tag{6.19}$$

in which, $v_1 = v(t - t')$
and, $\quad v_2 = v[t - \{t' + 2(t_d - t')\}] = v(t - 2t_d + t')$

$$\therefore v_{cr}(x, t) = K_{cr} \left[v_1 \left(t - \frac{x}{u} \right) - v_1 \left(t - 2t_d + \frac{x}{u} \right) \right] \tag{6.20}$$

In eqn (6.20) we have replaced t' by (x/u), the general variable x having been substituted for the particular value x'.

$\quad K_{cr}$ = reverse crosstalk coupling coefficient
$\quad\quad \triangleq \lambda u = \lambda/t_u$
or, $K_{cr} = [(C_M Z_0) + (L_M/Z_0)]/4t_u$ (6.21)

It can be shown that K_{cr} is dimensionless.
  A case of particular interest is $v_{cr}(0, t)$:

$$v_{cr}(0, t) = K_{cr}\{v_1(t) - v_1(t - 2t_d)\} \tag{6.22}$$

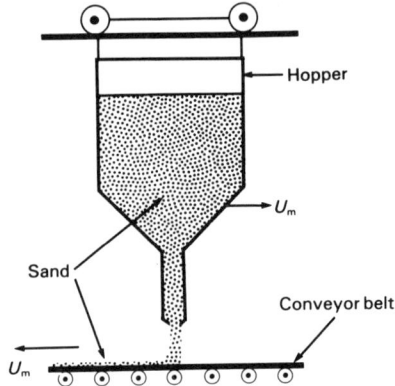

Fig. 6.7 Mechanical analogy for generation of reverse crosstalk waveform (compare with Fig. 6.5). The hopper and belt travel in *opposite* directions in this case.

For $t < (2t_d - x/u)$ is follows from eqn (6.20) that $v_{cr}(x, t)$ is an attenuated replica of $v_1(0, t)$.

We can use the hopper/conveyor belt set-up of Fig. 6.5 to illuminate the generation of reverse crosstalk. However, in this case the hopper moves forward at a velocity u_m while the conveyor belt moves at the same speed in the opposite direction (see Fig. 6.7). Sand is deposited on each part of the conveyor belt as it passes beneath the hopper exit-pipe and continues to arrive at the start (not shown) of the conveyor belt after the hopper has passed the end of the belt. Because the belt is moving backwards the depth of the layer of sand deposited on an incremental length, per-unit-time, is one half what it would be if the belt were stationary.

6.1.3   Crosstalk: calculation examples

Example 6.1

For the set-up of Fig. 6.2, $v_G(t)$ is the truncated-ramp waveform shown in Fig. 6.8. Sketch $v_{cf}(l, t)$ and $v_{cr}(0, t)$. Assume $t_c < 2t_d$, $K_{cf} < 0$, $K_{cr} > 0$.

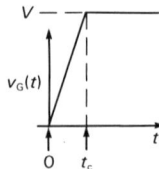

Fig. 6.8 $v_G(t)$ for Fig. 6.2, in Example 6.1.

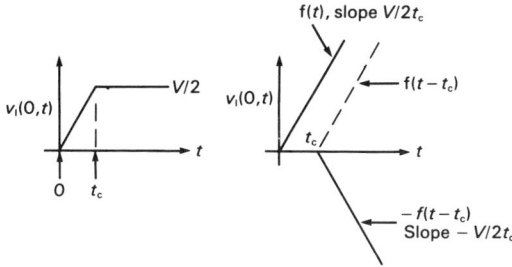

Fig. 6.9 Resolution of truncated-ramp drive waveform $v_1(0,t)$ on line A into a ramp and a delayed ramp with opposite slope.

Solution

The resolution of the truncated-ramp function drive waveform $v_1(0, t)$ on line A into a ramp waveform $f(t)$, having a slope $+ V/2t_c$, and a delayed ramp $f(t - t_c)$, with slope $- V/2t_c$, is shown in Fig. 6.9.

Thus, $v_1(0, t) = v_1(t) = \{f(t) - f(t - t_c)\}$ (6.23)

From eqn (6.14),

$$v_{cf}(l, t) = K_{cf} \, l \, \frac{\mathrm{d}}{\mathrm{d}t} \{v_1(t - t_d)\}$$

Replacing t in eqn (6.23) by $(t - t_d)$ gives,

$$v_1(t - t_d) = \{f(t - t_d) - f(t - t_d - t_c)\} \tag{6.24}$$

Figure 6.10(a) shows the waveform components $f(t - t_d)$, $f(t - t_d - t_c)$. From eqn (6.24),

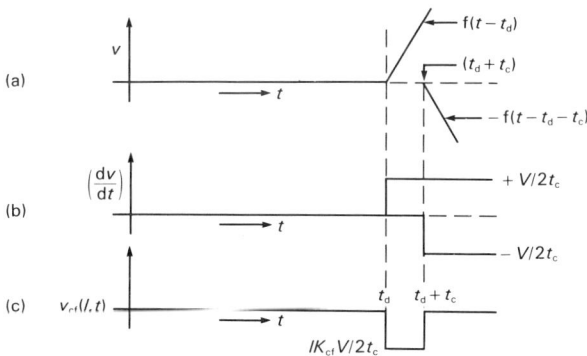

Fig. 6.10 Construction for forward crosstalk waveform $v_{cf}(l,t)$ for Example 6.1. (a) Showing $f(t - t_d)$, $-f(t - t_d - t_c)$; (b) Differentiated components of (a); (c) $v_{cf}(l, t)$ obtained from (b), assuming $K_{cf} < 0$.

$$\frac{d}{dt}\{v_1(t - t_d)\} = \frac{d}{dt}\{f(t - t_d) - f(t - t_d - t_c)\} \tag{6.25}$$

But, the result of differentiating a ramp of slope $+ V/2t_c$ is a step-function of magnitude $V/2t_c$. Consequently, eqn (6.25) is represented by the addition of the components in Fig. 6.10(b). Multiplying by K_{cf}, bearing in mind that $K_{cf} < 0$ (given), we obtain for $v_{cf}(l, t)$ the waveform shown in Fig. 6.10(c).

From eqn (6.22),

$$v_{cr}(0, t) = K_{cr}\{v_1(t) - v(t - 2t_d)\}$$

Using again our definition of f(t),

$$v_{cr}(0, t) = K_{cr}\{f(t) - f(t - t_c) - f(t - 2t_d) + f(t - t_c - 2t_d)\} \tag{6.26}$$

The graphical construction to find $v_{cr}(0, t)$ amounts to the superposition of a series of delayed ramp waveforms and is shown in Fig. 6.11.

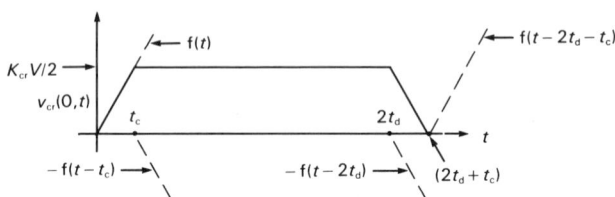

Fig. 6.11   Construction for obtaining reverse crosstalk waveform $v_{cr}(0, t)$
for Example 6.1.

The use of a truncated-ramp test waveform permits experimental estimation of K_{cf}, K_{cr} from observed waveforms such as those of Figs 6.10, 6.11.

A knowledge of the magnitudes of the coupling coefficients is then useful for estimating crosstalk with input drive signals of arbitrary shape. Although Example 6.1 specifies $t_c < 2t_d$, the graphical construction techniques used in Figs 6.10, 6.11 for finding $v_{cf}(l, t)$ and $v_{cr}(0, t)$ respectively are equally applicable to the case $t_c > 2t_d$ (see Problem P6.2).

The development of the topic of crosstalk and the evolution of the relevant equations, (6.13) and (6.20) has been based on the assumption that each line is correctly terminated at each end. However, this need not be the case. We can use the reflection chart, assuming the terminations are linear, to obtain modified waveforms. Of the various possibilities we consider only one, by way of illustration, that of Example 6.2.

Example 6.2

Sketch $v_{cf}(l, t)$, $v_{cr}(0, t)$ for the set-up in Fig. 6.2 if the termination Z_0 on line B at $x = l$ is replaced by a short-circuit. Assume, as in Example 6.1 that $v_G(t)$ has the form shown in Fig. 6.8, and also that $t_c < 2t_d$, $K_{cf} < 0$, $K_{cr} > 0$.

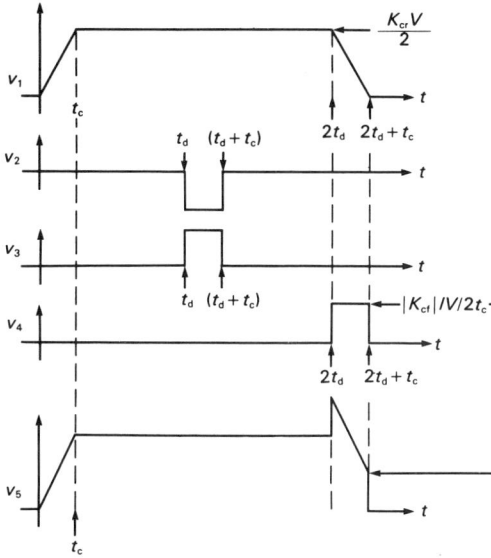

Fig. 6.12 Construction for obtaining $v_{cr}(0, t) = v_5$ for Example 6.2.

Solution

Refer to Fig. 6.12, where v_1 and v_2 are what would appear at $x = 0$ and $x = l$, respectively, on line B if it were matched at $x = l$. However, because of the short-circuit at $x = l$, $v_{cf}(l, t) = 0$, so v_3 (equal in magnitude but opposite in polarity to v_2) is generated at $x = l$. This propagates back along line B and appears as v_4. $v_5 = v_{cr}(0, t)$ is the algebraic sum of v_4 and v_1.

6.1.4 Reduction of crosstalk with microstrip lines

The equations we have developed for the 'ideal' case of two microstrip lines having top conductors that are parallel and close form the basis of some general observations that can be made to reduce unwelcome crosstalk effects.

For a given drive signal, v_1 and (dv_1/dt) are fixed. Hence, the following options are available to reduce crosstalk:

- Reduce the length of p.c.b. over which lines are parallel ($v_{cf} \propto l$). This may mean a modified layout for the p.c.b.
- Keep lines well spaced if, for some reason, they have to be parallel. This measure ensures that $|K_{cf}|$, $|K_{cr}|$ are minimized.
- Separate parallel lines by an interleaved third parallel line maintained at earth potential. The screening effect of this conductor reduces $|K_{cf}|$ and $|K_{cr}|$ further still.

6.2 CROSSTALK WITH CABLE BUNDLES: GENERAL COMMENTS

Interconnection between p.c.b.s must be made by some controlled impedance path, i.e. a transmission line. For neatness and reproducibility in system performance the transmission lines normally are grouped into cable bundles and it is possible that these will run parallel, and close, to each other over a distance of several inches.

Twisted-pair transmission line is frequently used in hard-wired systems. Crosstalk induced in one wire tends to cancel that induced in the other but nevertheless there is some residual crosstalk. This can be further reduced by twisting the pairs within a bundle, just as in plaitting hair. Twisted-pairs with an outer earthed screen offer increased immunity to crosstalk but are more expensive.

An attractive alternative to twisted-pair, particularly when p.c.b.s have to be removed from equipment for test or replacement, is the use of flat cable (connected to a p.c.b. edge connector). Flat cable is economical, neat and mechanically more flexible than twisted-pair. At its simplest flat (or ribbon) cable comprises a series of parallel conductors embedded in a plastic dielectric material. The whole entity, as its name suggests, has a planar cross-sectional geometry. Pairs of adjacent conductors can be used to form parallel two-wire lines and these can be screened from other pairs in the cable by earthing interleaved conductors.

The ultimate in crosstalk minimization is achieved by the use of miniature coaxial cable transmission lines. In these a signal conductor is completely enclosed in an outer earthed shield. Cost would normally rule out this solution for all but the most critical applications.

PROBLEMS

P6.1 Show directly from the definitions of K_{cf}, in eqn (6.9), and K_{cr}, in eqn (6.21), that K_{cf} has the dimensions time per-unit-length and K_{cf} is dimensionless.

> **Note:** In all the following problems assume the set-up of Fig. 6.2 with $K_{cr} = 0.015$ and $K_{cf} = -0.03$ ns/ft. t_d is specified, or implied, in each question. $t_u = 1.5$ ns/ft. For P6.2, P6.3, P6.4 assume $v_G(t)$ has the form shown in Fig. 6.8 with $V = 2V$, $t_c = 9$ ns.

P6.2 Sketch $v_{cr}(0, t)$ for (i) $t_c = 2t_d$; (ii) $t_c = 4t_d$.

P6.3 Sketch $v_{cr}(l/2, t)$ for $t_d = 30$ ns.

P6.4 Sketch $v_{cr}(0, t)$, $v_{cf}(l, t)$ if the termination at $x = l$ on line B is replaced by an open circuit ($t_d = 30$ ns).

> **Note:** For P6.5, P6.6, P6.7, assume $v_G(t)$ has the shape, and t_d the value, given in the problem.

P6.5 Sketch $v_{cr}(0, t)$, $v_{cf}(l, t)$ for $v_G(t) = 2[1 - \exp(-t/\tau)]$. Assume $\tau = 1$ ns, $t_d = 9$ ns.

P6.6 Sketch $v_{cr}(0, t)$, $v_{cf}(l, t)$ for $v_G(t) = V[1 + \tanh\{\alpha(t - t_h)\}]$, where $V = 1$ V; $\alpha = 2$ ns^{-1}; $t_h = 3$ ns; $t_d = 9$ ns.

P6.7 Sketch $v_{cr}(0, t)$, $v_{cf}(l, t)$ if v_G has the waveform of Fig. P6.7. ($t_d = 4.5$ ns)

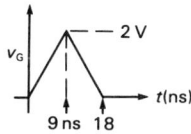

Fig. P6.7   Relating to problem P6.7.

7 Logic signal transmission: an introduction

An elementary digital system is shown in Fig. 7.1(a). It comprises a computer or a central processor unit (CPU) connected to a number of peripheral (P) units, P_1, P_2, P_3, etc., some, or all, of which may be located remotely with respect to the CPU. A common signal path, or data highway, links the CPU and P units. This is the *data bus* shown schematically as a single straight line with the P subsystems at the ends of short stubs.

The data bus is 'bidirectional'. Thus information can pass along it in either direction between the CPU and a P unit, or between P units. By virtue of the

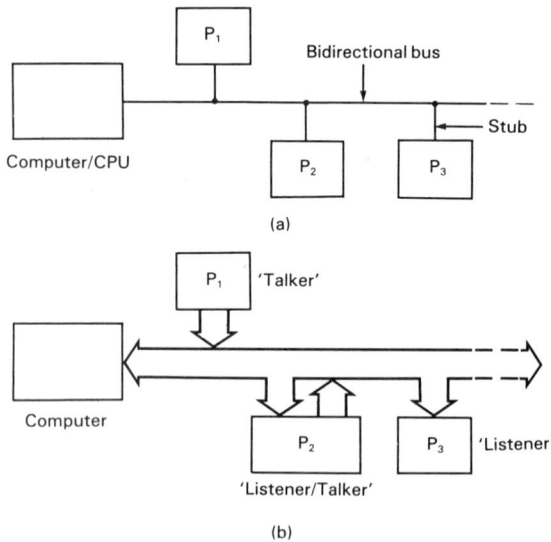

Fig. 7.1 Elementary digital system. (a) Schematic representation; (b) Showing data bus and one choice for peripherals.

two-way delay time of the bus in relation to the transition times of the digital signals used in modern systems, the bus must normally be regarded as a transmission line and terminated correctly to avoid undesired reflections.

A P unit can be a 'talker', a 'listener' or a 'listener/talker'. A talker is capable only of transmitting data; an example is a digital instrument outputting data. A listener is capable only of receiving signals, e.g. a printer. A 'listener/talker' (e.g. a calculator, or a memory) is capable of receiving data and, at a different time, transmitting data. An interesting case of a 'listener/talker' is a *modem* (*mo*dulator–*dem*odulator), which links a computer to a telephone network.

Figure 7.1(b) is a redrawn version of Fig. 7.1(a) for the case P_1 = talker, P_2 = listener/talker, P_3 = listener. The bus is shown now as a broad line and the arrows indicate the direction of possible signal flow.

We shall be dealing in this chapter only with the basic principles and practice associated with the hardware aspects of the bussing of digital signals, whether they are of a data or a control nature. Thus, we shall not concern ourselves with the necessary address and control aspects of information exchange, for example the selection mechanism for subsystems exchanging information, and the supervision required of the timing of the data transfer for the process to be accomplished satisfactorily.

Hardware design decisions that have to be made in the bus implementation of a digital system relate to line length, operating speed in Bauds (1 Baud = 1 bit/s), and noise environment. 'Noise' in this context includes: undesired reflected pulses occurring at line discontinuities resulting from the presence of stubs (see Chapter 5); line-to-line crosstalk, dealt with in Chapter 6; 'common-mode' noise due to the presence of mains-powered and other nearby equipment switching on and off, etc. and even (though, generally, less significant) electromagnetic interference from radio transmitters.

To avoid the inevitable confusion and inconvenience that would result from a multiplicity of equipment manufacturers' standards the Electronic Industries Association (EIA) has established a limited number of agreed standards of data transmission to cater for various physical bus lengths, system operating speeds and noise immunities. No attempt is made here to cover these in detail but two are outlined, by way of introduction to the subject, and their applicability to a system such as that shown in Fig. 7.1 discussed. Having grasped the fundamentals, the reader should be in a better position to appreciate the information contained in the data sheets of system component manufacturers (e.g. Texas Instruments, Signetics, etc.) and be able to profit more fully from their detailed application notes. First, however, it is necessary to set the scene by reviewing some essential characteristics of digital signal transmission.

7.1 TRANSMISSION CHARACTERISTICS

Three important features of digital logic signal transmission are considered here: parallel and serial data transfer; unbalanced and balanced drive; 'tristate' operation.

7.1.1   Parallel and serial data transfer

With parallel operation each binary digit (bit) of a word is allocated its own transmission path and time is the common feature. Thus, for a word of 8 bits (1 byte) each bit is available at the same instant of time. Fig. 7.2(a) shows the parallel representation of the 4-bit (1 'nibble') number 0110.

For serial operation the bits are transmitted in time sequence and a single transmission path is the common feature. Fig. 7.2(b) shows the serial representation of 0110.

The choice between parallel and serial transmission involves a trade-off between speed and hardware cost. In the example given, the serial mode of operation would be some four times slower than the parallel mode but would require four times less hardware.

'Mixed' operation gives the designer room for optimization. Thus, for fast internal data transfer a peripheral subsystem might operate in the parallel mode. However, to reduce external interconnection problems there could be a serial link to other peripherals. This approach involves the use of parallel/serial and serial/parallel converters but these are standard system components.

For simplicity, and because it embodies the basic properties we wish to study, we will consider only the case of serial data transmission. The next system choice is whether to use single-ended or differential line drive. As will be seen, this choice is governed by operating speed and noise problems.

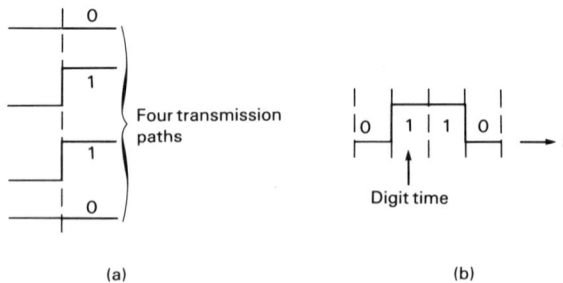

Fig. 7.2   Data transmission. (a) Parallel; (b) Serial.

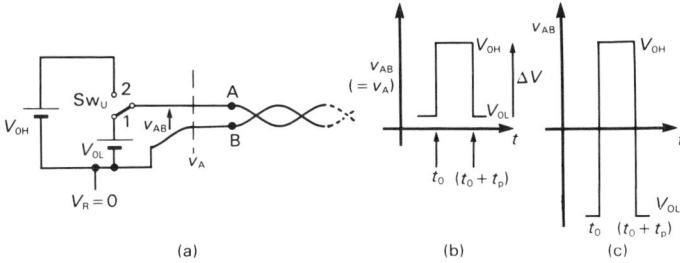

Fig. 7.3 Single-ended drive. (a) Equivalent circuit representation; (b) v_{AB} when condition of Sw_U changes momentarily; (c) v_{AB} for the case $V_{OL} = -V_{OH}$.

7.1.2 Unbalanced and balanced operation

A single-ended, or unbalanced, transmission circuit operates with one conductor held at a fixed reference potential, V_R, usually earth potential. A schematic representation is shown in Fig. 7.3(a). Points A, B define the inputs to a twisted-pair line, though a coaxial cable is equally applicable here. The circuit to the left of the dotted line simulates the action of a single-ended driver. V_{OH}, V_{OL} represent the logic levels '1', '0' respectively, used for transmission. When the wiper of the ideal single-pole, double-throw switch, Sw_U, moves from position 1 to 2 at time t_0 for a time duration t_p the resulting waveform v_{AB} is shown in Fig. 7.3(b). The drive is single-ended because the potential of B does not change.

For the special case $V_{OL} = -V_{OH}$, the extremes of voltage swing are balanced about earth potential but the drive circuit is, of course, still single-ended in nature. A particular case of single-ended drive, employing a 2-input TTL NAND gate, that gives v_{AB} similar to that in Fig. 7.3(b) is shown in Fig. 7.4.

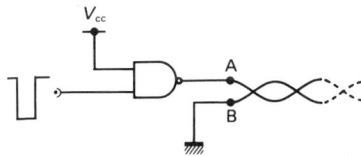

Fig. 7.4 (2-input) TTL gate used to drive a twisted-pair line.

In a differential, or balanced, transmission circuit the two conductors always have equal and opposite potentials with respect to a reference potential often conveniently, but not necessarily, earth potential. A schematic representation is shown in Fig. 7.5(a). Points A, B again define the

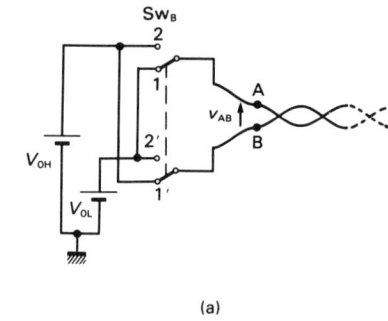

(a)

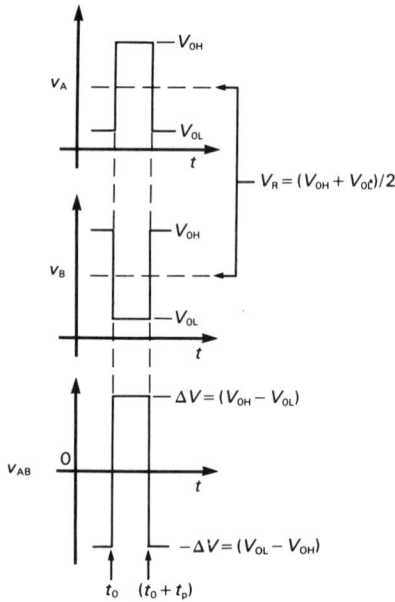

(b)

Fig. 7.5 Differential drive. (a) Equivalent circuit representation; (b) v_A, v_B, v_{AB} when condition of Sw_B changes momentarily.

input to a twisted-pair line and the circuit to the left of AB simulates the action of the balanced drive. When the (ganged) wipers of the ideal double-pole, double-throw switch Sw_B move from 1 to 2, and 1′ to 2′, at time t_0 for a time duration t_p, the resulting waveforms v_A, v_B, v_{AB} are shown in Fig. 7.5(b).

The signals v_A, v_B are balanced with respect to a reference potential $V_R = (V_{OH} + V_{OL})/2$. Conditions obtaining for $t < t_0$, $t > (t_0 + t_p)$ and for $(t_0 + t_p) \geqslant t \geqslant t_0$ are depicted by the equivalent circuits shown in Fig. 7.6(a), 7.6(b) respectively. We can write $v_{AB} = \pm \Delta V (\Delta V = V_{OH} - V_{OL})$, depending on the

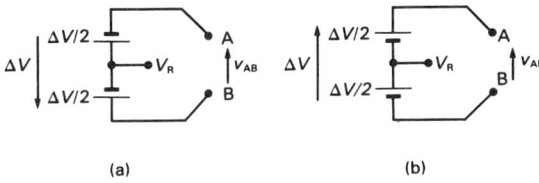

Fig. 7.6 Equivalent circuits for balanced drive: (a) for $t < t_0$ and $t > (t_0 + t_p)$; (b) for $(t_0 + t_p) \geqslant t \geqslant t_0$.

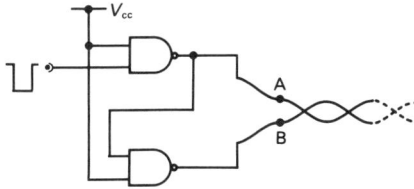

Fig. 7.7 Possible differential drive circuit using two 2-input TTL gates.

condition of Sw_B at a given time. The particular case $V_{OL} = -V_{OH}$ gives $V_R = 0$ and signals balanced about earth at both A and B.

Figure 7.7 shows how a differential drive circuit could, in principle, be implemented using standard TTL gates. It is apparent that to obtain differential drive a single rail supply can be used.

The relative merits of unbalanced and balanced drive are considered in section 7.2.

7.1.3 Tri-state operation

Bidirectional bussing at high data rates is made possible by the use of tri-state logic. As has been pointed out by Lewin (Ref. 1.7), the name is a misnomer as it does not involve a logic in which a variable can have three distinct values but the use of a third, open-circuit (high impedance) output state.

The operation of an unbalanced tri-state driver can be modelled by the simple circuit of Fig. 7.8. This is, basically, Fig. 7.3(a) with the addition of a

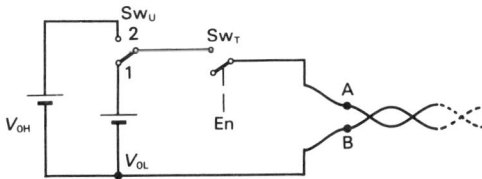

Fig. 7.8 Single-ended drive with tri-state facility.

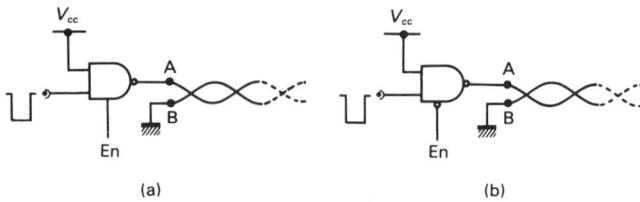

Fig. 7.9   TTL line drive with (a) 'Active-high' Enable and (b) 'Active-Low' Enable.

further single-pole, double-throw switch, Sw_T, incorporated for output isolation. When Sw_T is closed, operation is as described in section 7.1.2; when Sw_T is open a high-impedance state is seen looking back from the terminals AB. The state of Sw_T is governed by an 'Enable' control signal, *En*.

In the practical realization of Fig. 7.8 two choices exist, dependent on whether a high ('1') or low ('0') control signal causes Sw_T to be closed. The logic symbols for the two cases are given in Fig. 7.9 and Table· 7.1 summarizes their operation in relation to the circuit of Fig. 7.8.

A simple circuit model appropriate to the case of balanced drive with tri-state operation can be developed from Fig. 7.5(a); see problem P7.4.

Table 7.1
Showing how operation of the tri-state logic driver of Fig. 7.9 relate to Fig. 7.8.

		Logic level	Sw_T	Output
'Active-high' Enable,	En	H	Closed	Dependent on Sw_U
		L	Open	Z
'Active-low' Enable,	$\overline{En}$	H	Open	Z
		L	Closed	Dependent on Sw_U

H = High; L = Low; Z = High impedance.

7.2 NOICE REJECTION IN UNBALANCED AND BALANCED SYSTEMS

Both unbalanced and balanced techniques are used in logic data transmission.

Unbalanced operation has the advantages of low cost and simplicity but has the disadvantage of being practical only over 'short' distances with 'slow' transmission rates (and the associated slow logic waveform transition times). This is because it is easily affected by system noise.

Balanced operation, though somewhat more expensive than unbalanced

operation, is suitable – indeed mandatory – over 'long' distances and at 'high'
data transmission rates (and the associated fast logic waveform edges). This
is because it is far less susceptible to noise than unbalanced operation. The
words 'short', 'slow', etc. are shown in inverted commas because, as they
stand, they are purely qualitative and have no definite meaning in engineering
design. They are given more precise meanings in sections 7.3 and 7.4.

Consider first the case of unbalanced drive. It follows from Chapter 6 that
one form of noise (crosstalk) is less of a problem as a transmission path
decreases in length and logic waveform edges become slower. However, some
degree of crosstalk is an inevitable concomitant of digital system operation
and could be termed 'internal noise'. What of the effects of 'environmental'
noise? Voltage and current transients in equipment not specifically related to
the digital system can give rise to 'common-mode' noise, which can be a
serious problem with unbalanced operation.

The topic is not one that is easily dealt with analytically because the nature
of the noise source, the source impedance, the degree of coupling with the
transmission path, etc., are either poorly defined or even unknown quantities. However, a qualitative appreciation of the problem can be gained by
considering the particularly simple case of noise induced into a line at a
receiver input. In Fig. 7.10(a) the noise source NS is a relay being actuated.
There is the same coupling capacitance, C_N, between NS and each of the
conductors of the transmission path. A noise spike p (Fig. 7.10(b)) occurs at
A' because there is a finite impedance Z to earth at this input to the receiver R,
which is designed for a single-ended input. Z comprises the receiver input
impedance, Z_{IS}, in parallel with any line terminating resistor, R_T, that may be
used.

There is no significant noise spike produced at B', assuming there is a good
earth connection. If the amplitude of p – which is dependent on Z – is large
enough when A' is meant to be at a '0' level, then a spurious pulse occurs at
the output of R. A spurious pulse also occurs at the output of R if a negative-

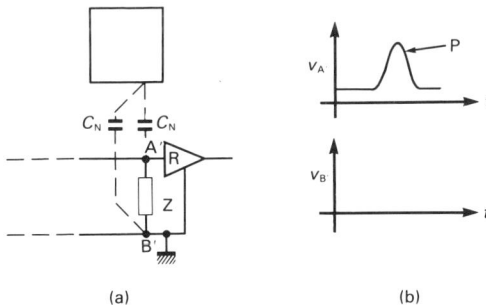

(a)   (b)

Fig. 7.10   Injection of noise spikes into an unbalanced transmission path
by a noise source NS. (a) Simple system model; (b) Waveforms at A' and
B'. p is a noise spike.

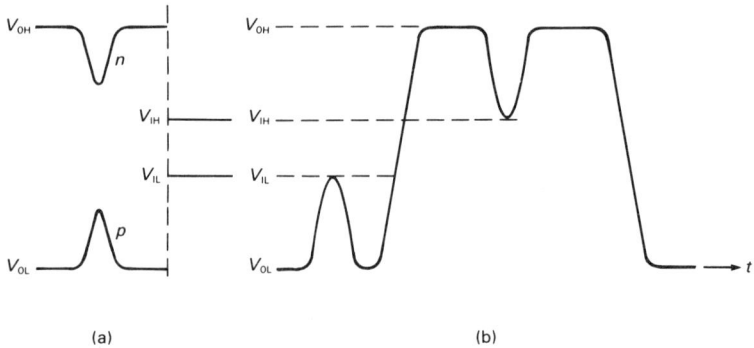

(a)   (b)

Fig. 7.11 A receiver is 'blind' to noise spikes when noise margins are sufficient, whether for spikes on quiescent levels, (a) or associated with a data pulse, (b).

going noise spike, n, of sufficient amplitude is produced at A' when it is at the '1' level.

Figure 7.11(a) shows that if the receiver input logic '0' level V_{IL} is sufficiently positive with respect to the driver output '0' logic level V_{OL}, then the receiver will be 'blind' to p and there will be no spurious output. Similarly, if the receiver input '1' level V_{IH} is sufficiently negative with respect to the driver output '1' level V_{OH} then the receiver will also be 'blind' to n. What has been said about spikes on quiescent levels applies equally well to spikes occurring while data pulses are being transmitted. In Fig. 7.11(b) the maximum amplitude for p is $(V_{IL} - V_{OL})$ and the maximum amplitude of n is $(V_{OH} - V_{IH})$.

If an unbalanced transmission path is badly corrupted by noise we may even have the state of affairs shown in Fig. 7.12 where the spikes invade the

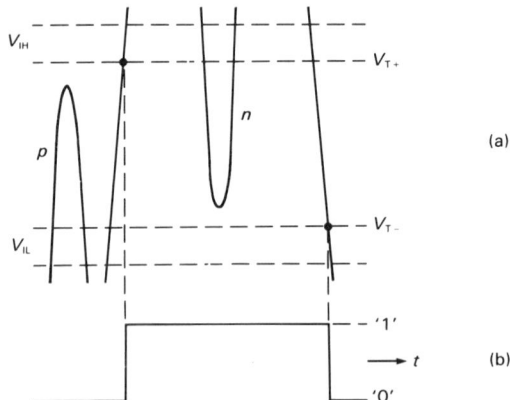

Fig. 7.12 Provision of extra noise immunity by use of waveform-edge, polarity-sensitive, switching levels. (a) Noise-contaminated waveform; (b) Receiver output waveform.

region between V_{IL} and V_{IH} that is forbidden except for transitions between logic levels. However, all is not lost. There is a circuit (the 'Schmitt trigger') that gives its name to a receiver transfer function that is capable of achieving the seemingly magical task of eliminating both p and n in Fig. 7.12. It does this by introducing switching threshold levels V_{T+} ($< V_{IH}$) and V_{T-} ($> V_{IL}$).

Which one of these two is operative at a given time depends on the polarity of the slope of the input waveform. V_{T+} exists only for a positive-going waveform edge, i.e. $(dv_I/dt) > 0$, and V_{T-} exists only for a negative-going waveform edge, i.e. $(dv_I/dt) < 0$. Thus, with the noise spikes shown in Fig. 7.12(a) the receiver output waveform is 'clean' as shown in Fig. 7.12(b). The transfer functions for the receiver of Fig. 7.13(a) are shown in Fig. 7.13(b), 7.13(c), for v_I increasing and decreasing, respectively. ($V_{T+} - V_{T-}$) is known as the 'hysteresis' voltage.

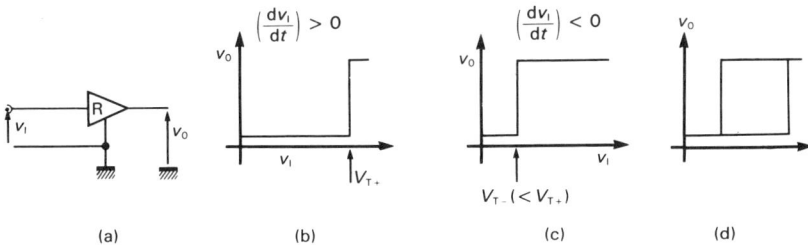

Fig. 7.13 A single-ended receiver with hysteresis for added noise rejection. (a) Receiver symbol; (b) v_0 v. v_I for increasing v_I; (c) v_0 v. v_I for decreasing v_I; (d) Hysteresis symbol.

A hysteresis symbol, sometimes put inside the receiver triangular symbol when applicable, is shown in Fig. 7.13(d). It takes the form of an idealized hysteresis loop, met with in the study of the properties of magnetic materials, and is actually a combination of Fig. 7.13(b) and (c). Some practical receivers have a facility for hysteresis adjustment at an external terminal.

Unbalanced operation is quite incapable of dealing with noise pulses that have amplitudes in excess of data logic levels. If these noise pulses are due to common-mode effects, then balanced operation is obligatory. Consider Fig. 7.14. In this a differential receiver is employed. Z represents the differential input impedance in parallel with any line terminating resistor R_T. Z_{CM} represents a common-mode impedance to earth at each receiver input terminal. In this case identical noise spikes p and p' (Fig. 7.14(b)) occur at the points A' and B' respectively because the impedances to earth at these points are identical. There is no net change in the receiver differential input voltage, $V_{A'B'}$, and no spurious output from R.

Suppose, now, that a data pulse is sent along the balanced transmission path. It is apparent from Fig. 7.15 that there is a 'clean' input $V_{A'B'}$ to R despite the presence of common-mode noise spikes p, p' and n, n' which may

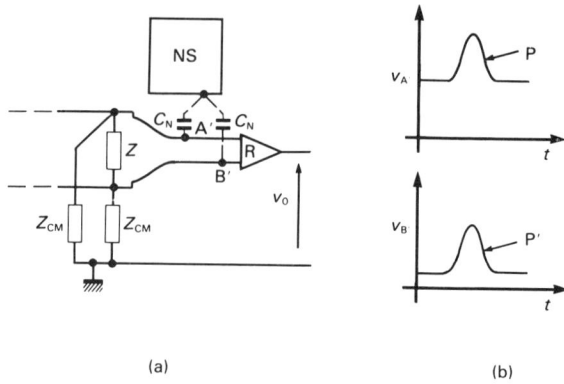

(a)   (b)

Fig. 7.14 Injection of noise spikes into a balanced transmission path. (a) Simple system model; (b) Waveforms at A′ and B′. p and p′ are identical in time, amplitude and shape.

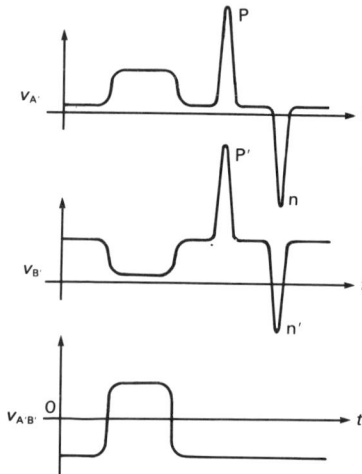

Fig. 7.15 Showing the excellent common-mode, noise-rejection characteristics of a differential receiver.

far exceed the amplitude of the data pulse. Thus, environmental noise has far less effect on balanced-mode operation than unbalanced operation. It is possible to 'rescue' a data signal 'buried' in noise. Reference to Figs 7.5 and 7.6 may also help to demonstrate this: V_R may be regarded as a common-mode signal, which does not affect v_{AB}.

Because only a small logic swing is necessary even in the presence of appreciable common-mode noise, balanced operation permits data transmission at high rates in a noisy environment. Of course, a high rejection of common-mode signals does not help eliminate *differential-mode* noise pulses that may arise on a line due to reflections at discontinuities at points where the stubs join. Noise margins for reflected pulses can be increased by using differential receivers with in-built hysteresis.

7.3 SINGLE-ENDED POINT-TO-POINT TRANSMISSION: RS232C

Figure 7.16 shows a single-ended point-to-point transmission link. Data flow is undirectional from D to R. Depending on the system (e.g. Teletype machine link) it may be necessary to implement this with the RS232C, a popular and still widely used variation of the first IEA standard on data transmission, RS232, introduced in 1962. An RS232C driver gives a bipolar output signal, necessary for use with the polarized magnets of teleprinters in peripheral units.

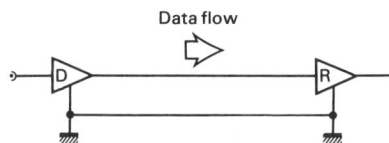

Fig. 7.16 Single-ended point-to-point transmission.

Abbreviated operating data, relevant to the hardware, are as follows (See Ref. 7.2):

Driver ouput signal	± 5 V (min.)
	± 15 V (max.)
Driver load	3 kΩ–7 kΩ
Driver slew rate	30 V/μs (max.)
Data transmission rate	20 kB (max.)
Cable length (l)	15 m (max.)
Receiver sensitivity	± 3 V
Receiver input resistance	3 kΩ–7 kΩ

There are many commercially available drivers and receivers able to meet these requirements. (Note: A normal 5 V TTL gate is unsuitable for D because of the bipolar nature of the output signal.)

RS232C illustrates the significance of the words 'short' and 'slow' mentioned in section 7.2. With a maximum cable length of 15 m and a twisted-pair delay time t_u of, say, 7 ns/m the two-way delay time $2t_d$ is less than 250 ns. 'Slew-rate' defines speed of waveform edge. At 30 V/μs, the receiver input range -3 V to $+3$ V is traversed in a minimum of 200 ns. From the discussion of Chapter 1 the system is on the limit of being described as 'distributed'. Pulse reflections are likely to affect the pulse edges (see Chapter 3, Fig. 3.14).

A more recent IEA standard for single-ended transmission, RS423, provides for up to ten receivers driven by one driver, line lengths up to 1200 m and data transmission rates up to 100 kB. More detailed information is given in Ref. 7.2.

7.4 MULTI-POINT BIDIRECTIONAL DATA TRANSMISSION: RS485

To operate up to 1 km with data rates of 100 kB requires balanced operation. Bidirectional bussing requires a driver that can be switched into a high-impedance output state. RS485 is an IEA standard applicable to 'high' speed multi-point data bussing, or party-line systems. Balanced operation is employed but bipolar signals are not mandatory.

Abbreviated data, relevant to the hardware are as follows:

Number of driver/receiver pairs	32
Driver output signal	± 1.5 V (min.)
Driver load	60 Ω
Cable length (l)	1200 m (max.)
Data transmission rate	10 MB for $l = 12$ m
	1 MB for $l = 120$ m
	100 kB for $l = 1200$ m
Receiver sensitivity	± 300 mV
Receiver input resistance	> 12 kΩ

Two features deserve special mention: the 60 Ω load capability means that a 120 Ω twisted-pair line can be terminated at each end, a desirable feature in a bidirectional bus; the trade-off between data transmission rate and line length results from the increase in line attenuation with frequency (see Appendix A).

Figure 7.17 summarizes the description of a tri-state transmitter and tri-state receiver for use with RS485. The transmitter and receiver may be contained in one package as a 'transceiver'. A transceiver, symbolized by a rectangular outline, is shown in Fig. 7.18 driving a 120 Ω twisted-pair bus terminated at both ends in its characteristic resistance. If DE and $\overline{RE}$ are both 'high', then the transceiver acts as a driver and the data signal at D appears in differential drive form at A and B. However, if DE and $\overline{RE}$ are both 'low' then the transceiver acts as a receiver and the differential data signal appearing at A and B is converted to a single-ended TTL-logic compatible signal at R.

By the use of appropriate transmitters and receivers the simple digital system of Fig. 7.1 is readily implemented.

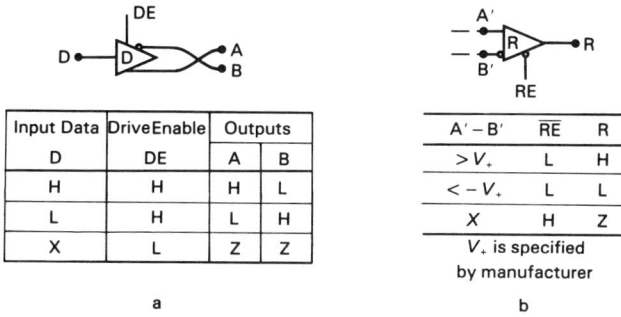

Input Data	DriveEnable	Outputs	
D	DE	A	B
H	H	H	L
L	H	L	H
X	L	Z	Z

A' – B'	RE	R
> V+	L	H
< – V+	L	L
X	H	Z
V+ is specified		
by manufacturer		

a   b

Fig. 7.17 Balanced driver/transmitter (a) and balanced receiver, (b), showing symbolic representation and logic operation. H = High; L = Low; X = Don't care; Z = High impedance. For the SN75176A transceiver, $V_+ = 0.2$ V.

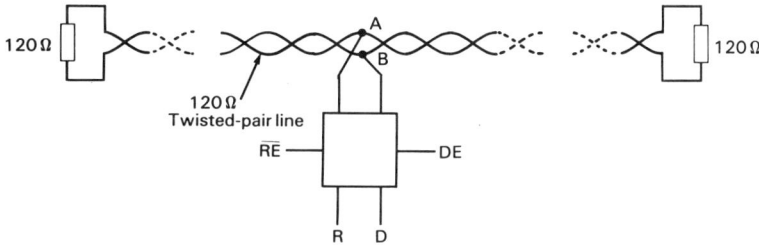

Fig. 7.18 One of 32 possible transceivers connected to a line, matched at both ends, in an RS485 link.

PROBLEMS

P7.1 In Fig. P7.1, Cable 1 is to be matched at terminal X and Cable 2 at terminal Y. Calculate R_1, R_2, R_3 if $(v_Y/v_X) = 0.5$, assuming each cable has a characteristic impedance of 50 Ω.

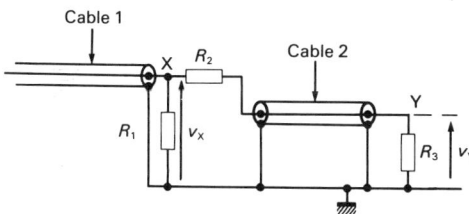

Fig. P7.1 Relating to problem P7.1.

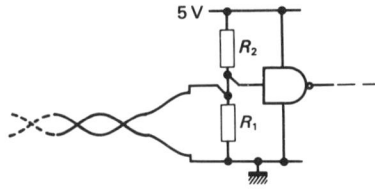

Fig. P7.2 Relating to problem P7.2.

P7.2 A common-emitter saturating switch drives a length of twisted-pair
(characteristic impedance 100 Ω), which is terminated by R_1, R_2 at the
input to a TTL logic gate, as shown in Fig. P7.2.
Determine a pair of suitable values for R_1, R_2, in a preferred resistor
range, if the line is to be matched and the gate input voltage at the '1'
level is to exceed 3 V.
What constraint does the choice of R_2 place on the line drive circuit?

P7.3 Fig. P7.3 shows a possible emitter-follower line drive circuit driven by
a TTL gate. Comment, briefly, on desirable device parameters for the
transistor Q in this application.
Why would the emitter-follower drive stage, shown, be inappropriate
as a high-impedance probe for the detailed observation of fast digital
signals?

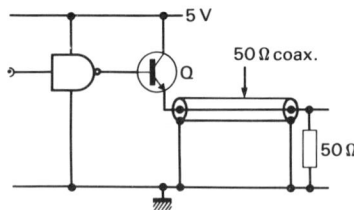

Fig. P7.3 Relating to problem P7.3.

P7.4 Draw a simple circuit model, similar to that of Fig. 7.5(a), to simulate
the operation of a balanced drive circuit with a tri-state output
facility.

P7.5 The impedance seen at x, x', in Fig. P7.5, looking back into each

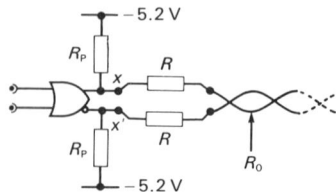

Fig. P7.5 Relating to problem P7.5.

output of the ECL gate is approximately 10 Ω. Specify a suitable preferred resistor value for R if the 100 Ω twisted-pair line is to be matched at its sending end.

P7.6 For the balanced drive circuit in Fig. P7.6(a), $V_{OH} = 2.8$ V and $V_{OL} = 0.4$ V. Calculate the current required of the driver stage.

What is the maximum drive current required when the resistor R_T in Fig. P7.6(a) is replaced by the arrangement shown in Fig. P7.6(b)?

In what respect may the use of the termination shown in (b) be preferable to that shown in (a)?

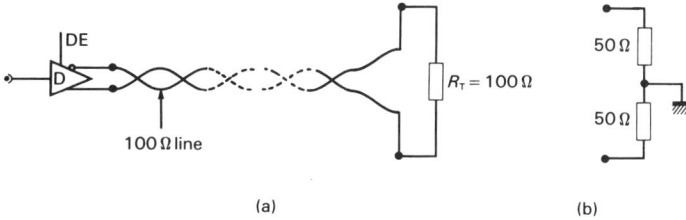

(a)   (b)

Fig. P7.6   Relating to problem P7.6.

P7.7 Give a sketch showing how two drivers and two receivers with the characteristics depicted in Fig. 7.17 can be interconnected to function as a bidirectional line repeater.

P7.8 Figure P7.8 shows a 100 Ω twisted-pair line matched at its sending and receiving ends. It is driven by a current-mode switch, modelled by the components inside the dashed box, often employed in high-speed digital data communication systems. The current I flows to the line via contact 2 or contact 1 of the switch, Sw, dependent upon the logic level of an input data signal, not shown. The transition time of the switch can be taken as zero (in practice 10 ns is the order of magnitude possible).

Calculate the magnitude of I, in mA, if the differential line voltage between the terminals A, B, when the switching action occurs, is to be 1 V.

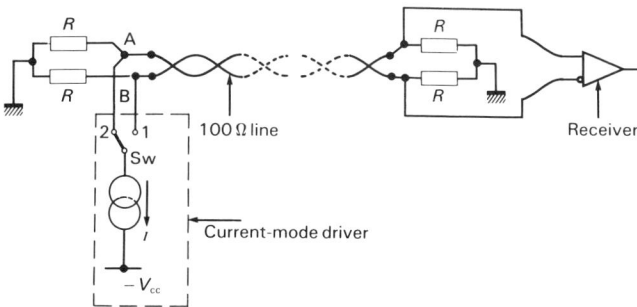

Fig. P7.8   Relating to problem P7.8.

Appendix A: résumé of practical line characteristics

In Chapter 2 we defined the parameters and properties of an ideal lossless line, and subsequent analyses and discussion were based on that ideal concept.

Figure A.1 summarizes the structure, characteristic impedance and application potentialities of some practical lines. Some points of general interest are:

- Cases (a) and (e), only, are 'balanced' lines.
- Z_0 is only weakly dependent on the accuracy of mechanical dimensions because of the logarithmic dependence. Thus, if (d/a) increases by an order of magnitude, $\log_e(d/a)$ increases only by a factor of 2.3.
- Z_0, t_u, for any of the lines shown, can be found quickly using a TDR technique. For the determination of Z_0 low-inductance metal film resistors can be used as test terminations.

Attenuation has been ignored, since lossless lines have been assumed. However, attenuation though often small, is finite in practice. Thus for the cable type UR-67 (50 Ω, polyethylene) one manufacturer quotes an attenuation of some 2 dB per 100 ft at an operating frequency of 100 MHz. This means a reduction in signal amplitude of about 20% at that frequency, over a distance of 100 ft. The effects of attenuation in fast pulse work are minimized by choosing a low-loss cable and interposing 'repeaters', if the signal level has become degraded, at appropriate points when long cable runs are used.

For analogue signals a repeater must be a low-noise, wide-band amplifier: for a digital logic signal a repeater is a 'reshaping' or 'regenerator' circuit that restores pulse rise and fall times and logic levels (see Chapter 7).

Line type	Physical form	Formula for $Z_0(\Omega)$*	Comments/applications
(a) Parallel wire	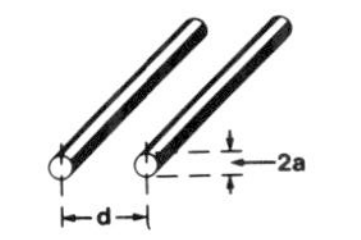	$(120/\sqrt{\epsilon_r})\log_e(d/a)$	Not normally used, as such, in high-speed electronics as other types of line are more appropriate. However, it is used in flat, ribbon, cable (see Chapter 6).
(b) Wire-over-ground	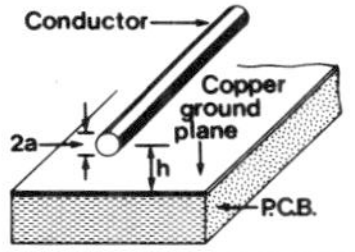	$(60/\sqrt{\epsilon_r})\log_e(2h/a)$	Unavoidable when components supported over chassis-ground of p.c.b. copper surface. Can be exploited in high-speed design work.
(c) Microstrip	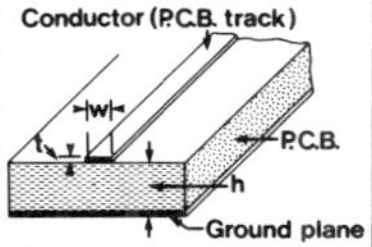	$\{87/\sqrt{(\epsilon_r + 1.41)}\}$ $\log_e[5.98h/(0.8w + t)]$	Widely used with p.c.b.s.
(d) Coaxial cable ('Coax')	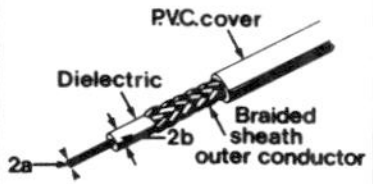	$(60/\sqrt{\epsilon_r})\log_e(b/a)$	Best for pulse transmission but bulky, though miniature cables available. $Z_0 = 50\ \Omega$ is an international standard in pulse electronics.
(e) Twisted-pair		$\approx (120/\sqrt{\epsilon_r})\log_e(D/a)$ D = centre-to-centre distance between wire centres: a = wire radius. Typically, Z_0 ranges from 100–120 Ω	Gauge of wire (typically 26 gauge, 0.46 mm) and number of twists/metre do not critically affect Z_0†. Easy to construct and cheap. Inter-p.c.b. substitute for co-ax. Screened twisted-pair commercially available.

*ϵ_r = Relative permittivity: for polyethylene $\epsilon_r \approx 2.3$ and for glass epoxy $\epsilon_r \approx 5$.
$t_u = (1/u) \approx 3.3\sqrt{\epsilon_r}$ ns/m: for polyethylene 50 Ω coax., $t_u \approx 5$ ns/m.
†However, t_u increases with number of twists per metre.

Fig. A.1 Summary of some practical line properties.

Appendix B: laboratory demonstration work

This appendix is written for those readers who wish to acquaint themselves, in the laboratory, with the practical aspects of transmission-line pulse-electronics but who do not have ready access to very wide-band oscilloscopes (either real-time or sampling) and TDRs.

GENERAL CONSIDERATIONS

Consider the set-up in Fig. B.1. PG is a pulse generator, with output impedance Z_G, that supplies either a triggered or a free-running rectangular waveform to a transmission line in the form of a coaxial cable of characteristic impedance R_0 and one-way delay-time t_d. PG also supplies a trigger pulse to the cathode ray oscilloscope (CRO). Waveforms $v_I(t)$, $v_T(t)$ are displayed on the Y_1, Y_2 channels respectively.

Let the 10%–90% rise-times of the pulse generator, cable, and oscilloscope

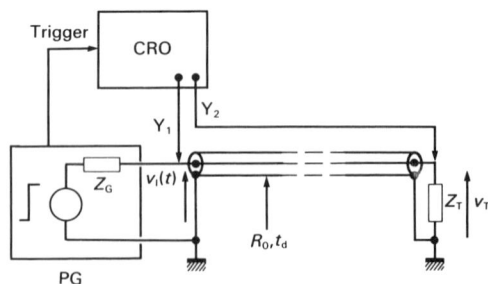

Fig. B.1 Set-up for basic practical work.

be t_{pr}, t_{cr}, t_{or} respectively. If the oscilloscope is to exhibit distinctive horizontal sections of waveform on the Y_1 channel, and thus illustrate the distributed nature of a transmission line as compared with a lumped circuit, for $Z_T \neq R_0$ it is necessary that:

$$2t_d > \sqrt{(t_{pr}^2 + t_{cr}^2 + t_{or}^2)} \qquad (B.1)$$

Note that t_{cr} has been ignored in all our discussions because we considered only lossless lines. However transmission lines do have a finite, though small, t_{cr}; the interested reader is referred to the book by Matick listed in the References. The 50 Ω cable type RG-58U has $t_{cr} < 1$ ns per 10 ft length.

As yet, t_{pr} is unspecified but we can make an estimate of a possible value, e.g. ECL, which employs the long-tailed pair, switches well within 4 ns.

Laboratory oscilloscopes with a bandwidth in excess of 30 MHz ($t_{or} \approx 12$ ns) tend to become expensive. It is thus evident that t_{or} has a dominant effect in the determination of the minimum acceptable t_d.

Using the figures quoted, we require:

$$t_d > 6.5 \text{ ns} \qquad (B.2)$$

Allowing a cable delay of 5 ns/m, inequality eqn (B.2) is easily met with, say, 2.5 m of cable which is not an impractical length.

It should be appreciated that the demands on t_{or} are most easily satisfied by using, say, 50 yards of cable on a reel. However, long lengths such as this introduce the problem of interpreting waveforms distorted by cable losses, which is best left until the case of the 'loss-free' cable has been understood.

CIRCUIT DESCRIPTIONS AND OBSERVED WAVEFORMS

A block schematic for the cable drive circuit is shown, between the dotted lines, in Fig. B.2. A pulse from virtually any low-cost, rectangular-waveform generator (including one based on a simple '555 timer' astable circuit), is squared-up by passing it through two TTL stages, TTL1, TTL2 connected in series. (If the waveform generator has a TTL output, TTL1, TTL2 are unnecessary but if a sinusoidal signal source, only, is available its output can be squared-up by using a TTL Schmitt trigger.)

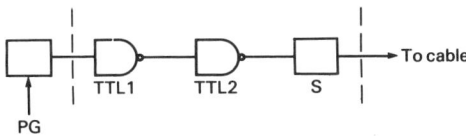

Fig. B.2 Block diagram of cable drive arrangement.

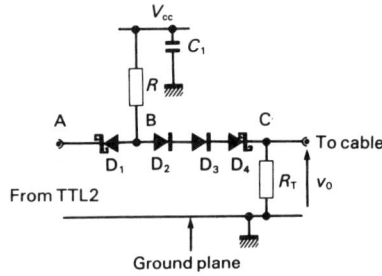

Fig. B.3   Simple form for S of Fig. B.2. D_1, D_4 = Schottky diodes; D_2, D_3 = IN4148; R = 1 kΩ (metal film); R_0 = 50 Ω BNC termination; C_1 = Ceramic capacitor (e.g. 1000 pF) for rail supply decoupling.

The output of TTL2 is not suitable for driving a cable directly for two reasons. First, pulse rise-time is relatively large (≈ 10 ns); second, the output impedance of a TTL stage is not known with certainty and varies over the output voltage range. Pulse shaper S overcomes these difficulties. Rise-time is reduced by 'slicing' a section from TTL2's output. A known source impedance is achieved by arranging for S to drive a known current into a cable with a specified resistor connected across its output terminals.

One form for S is shown in Fig. B.3. The 'earth' should be a printed circuit board copper ground plane. The p.d. across R_T is zero when A is at logic '0'. When A is at logic level '1', point C assumes a potential v_0, where

$$v_0 = [R_T/(R_T + R)][V_{cc} - \Sigma v_D] \tag{B.3}$$

Σv_D is the sum of the voltage drops across conducting diodes D_2, D_3, D_4. Variation of V_{cc} gives amplitude control. Equation (B.3) is applicable provided, under all conditions,

$$+ 1 \text{ V} > v_0 > -0.5 \text{ V} \tag{B.4}$$

Figure B.4 shows the delay-line 'differentiation' that results when a length of cable (RG213U), shorted at the far end, is connected to C. (For this particular test a wide-band oscilloscope *is* necessary.)

In cases where it is desirable to use a fixed voltage supply V_{cc}, the resistor R of Fig. B.3 may be replaced by a variable current source such as a suitably biased common-base transistor stage with a potentiometer in its emitter circuit. Thus, in Fig. B.5, which shows an improved version of the shaper S of Fig. B.3, V_{cc} and R_B are chosen so that zener diode D_z operates in the breakdown region. Q, operating in the common-base mode, supplies a collector current I from a high incremental source impedance and adjustment of R_V permits variation in the magnitude of I, which is given by

$$I \approx V_z/(R_v + R_E) \tag{B.5}$$

where V_z in the p.d. across D_z in the breakdown region.

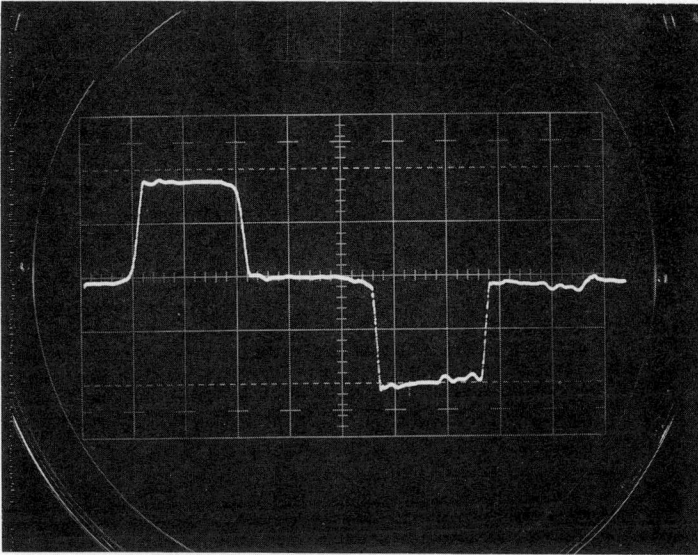

Fig. B.4   Waveform at C of Fig. B.3 with a line shorted at the far end
connected to C. Vertical scale 100 mV/cm; horizontal scale 20 ns/cm.

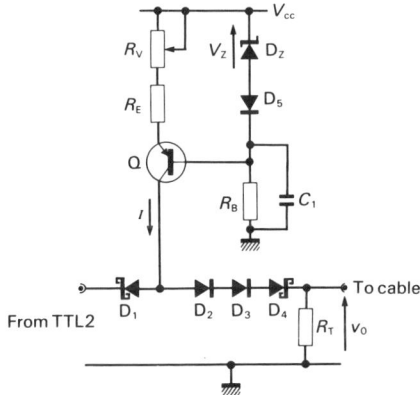

Fig. B.5   Alternative form for S of Fig. B.2. D_2, D_3, D_5 = IN4148; D_1, D_4
= Schottky diodes; D_Z = IN751A; Q = Any pnp with $C_{ob} < 3$ pf; C_1 =
1000 pF Ceramicon; R_T = 50 Ω BNC termination; R_v = 1 kΩ; R_E = 220 Ω;
R_B = 1 kΩ; $V_{cc} \geqslant 10$ V.

I is sensibly constant even when V_{cc} is not well stabilized. When the output
of TTL2 is at '0' it acts as a current sink for I; when its output is at '1', I
divides between R_0 and the cable. As with the simple circuit of Fig. B.3, the
sending-end impedance for any pulses reflected from the far end of the cable
($Z_T \neq R_0$) is R_0 provided the condition (B.4) is satisfied.

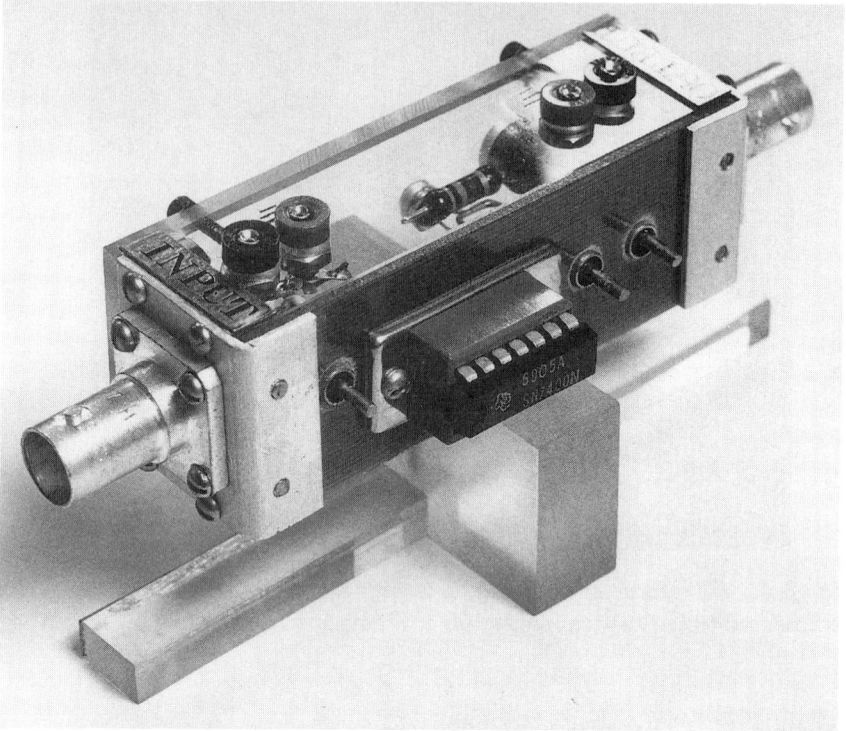

Fig. B.6   View of practical cable drive unit.

Note that Q in Fig. B.5 need not be a high-speed transistor because it never switches, but it must have a low C_{ob} in order not to degrade, unduly, the output pulse rise-time. For operation with a 5 V rail, D_Z can be replaced by a band-gap voltage reference source (e.g. Intersil ICL 8069, which has a breakdown voltage of 1.23 V approx.), provided R_V, R_E are reduced in value.

A practical version of that section of Fig. B.2, between the dotted lines, that includes the shaper S of Fig. B.5 is shown in Fig. B.6. For mechanical rigidity and protection of components from dust and damage the assembly consists basically of a box approx. 3 in × 1 in × 1 in (76 × 25 × 25 mm), one side of which is a p.c.b. with copper covering its inside surface to form a ground plane. Aluminium brackets at right angles to the board constitute the ends and in these are mounted BNC 50 Ω panel sockets for input and output cables. Holes in the board permit the outer cases of low-inductance ceramicon decoupling capacitors to be soldered directly to the ground plane. Another hole accommodates a DIL socket for removal of the (TTL) IC package.

The internal construction and wiring are visible if the remaining sides are made of transparent plastic pieces screwed together. Holes allow R_V to be adjusted externally and accommodate small sockets for power leads. An (optional) plinth ensures that the unit stands sufficiently far above a bench to allow the cable sockets to be easily accessible.

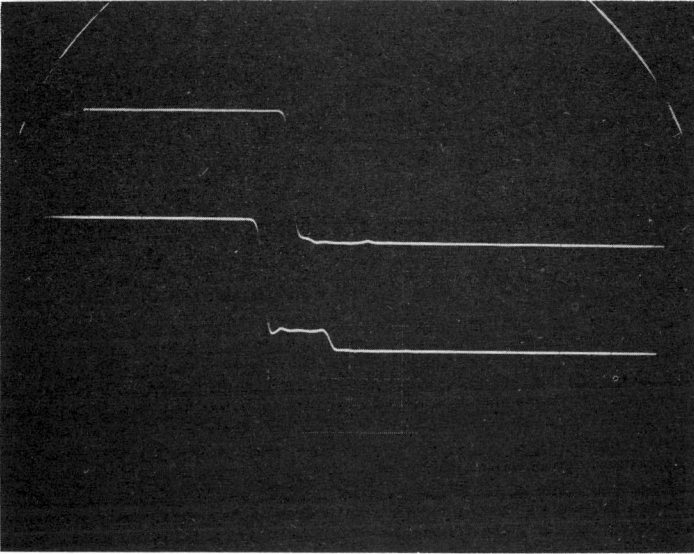

Fig. B.7 Waveforms for $v_I(t)$, upper trace, and $v_T(t)$, lower trace, for the scheme of Fig. B.1 using the unit in Fig. B.6. $Z_G = R_0 = 50 \, \Omega$; cable has $R_0 = 50 \, \Omega$; $Z_T = R_T = 75 \, \Omega$. Horizontal scale: 40 ns/major division; vertical scale: 50 mV/major division.

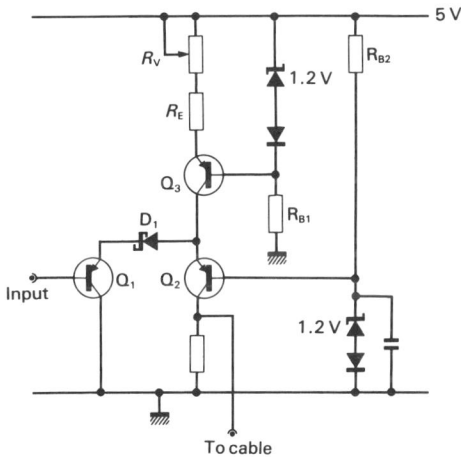

Fig. B.8 Further, improved, form for S of Fig. B.2.

Figure B.7 shows waveforms obtained using the unit of Fig. B.6. Many improved designs for the circuit of Fig. B.5 exist. Thus, Fig. B.8 shows, in outline only, a scheme that offers much greater isolation between the input and output circuits. It should, like the circuit of Fig. B.5, use a p.c.b. with a copper ground plane.

In Fig. B.8, Q_3 gives a collector current I that is switched by the long-tailed pair comprising the high-speed transistors Q_1, Q_2 (e.g. 2N4260). The conducting state of Q_1, Q_2 depends on the logic level at the base of Q_1. D_1 prevents the possibility of reverse breakdown in Q_1 should large amplitude inputs be applied.

Appendix C: general notes on waveform observation

The observation of waveforms with 'fast' edges, e.g. ECL signals, can present problems to the less experienced engineer involved with pulse and digital system hardware. The aim of this appendix is to offer some practical hints to help the reader avoid some of the more obvious pitfalls.

In order to ease the transition from the familiar to the less familiar, we review first the observation of waveforms in lumped systems. We then consider the special precautions necessary for waveform monitoring with distributed systems.

LUMPED SYSTEMS

Most basic laboratory oscilloscopes (CROs) with a bandwidth in the region of 30 MHz (and a corresponding rise-time of about 12 ns), have BNC input connectors for use with coaxial leads, the outer-conducting sheaths of which are earthed when connected to the instruments. These CRO leads thus offer good screening against unwanted, but often unavoidable, electromagnetic interference that can be a problem particularly with low-level signals.

Suppose it is required to monitor a waveform having a minimum transition time, t_c, of, say, 200 ns appearing across a resistor R_G, one side of which is earthed in the circuit under test. The typical set-up shown in Fig. C.1(a) could be used. The CRO lead is assumed to have a length of approximately 1 m – corresponding to a two-way delay time $2t_d$ of about 10 ns – and a total capacitance, C_T, of the order of 90 pF. The CRO input resistance, R_I, and input capacitance, C_I, tend to be standardized by equipment manufacturers at 1 MΩ and about 30 pF, respectively.

The measurement system can, in this case, be regarded as lumped (see

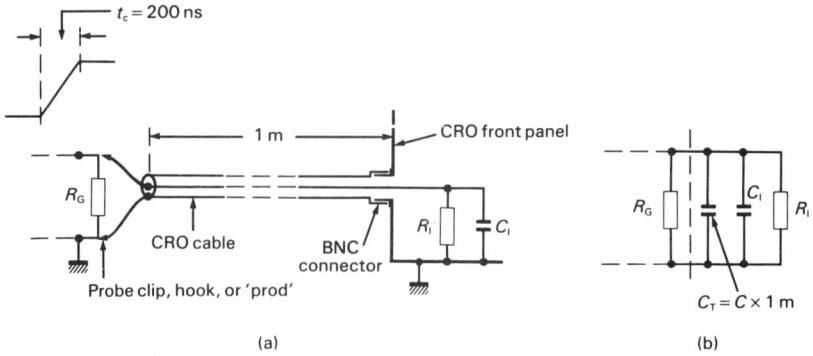

(a)

(b)

Fig. C.1 (a) Acceptable measurement scheme for $t_c \gg 2t_d$; (b) Section to right of broken line shows equivalent loading on R_G. $R_I = 1$ MΩ, $C_I \approx 30$ pF.

Chapter 1), as $2t_d \ll t_c$, and the equivalent loading across R_G presented by the lead and oscilloscope, is shown to the right of the broken line in Fig. C.1(b). If the magnitude of either the resistive or capacitive loading across R_G is unacceptable, then we can make use of the long-established and popular 10 × ('10 times') probe, one version of which is shown schematically in Fig. C.2(a). The insulated probe-housing provides a facility allowing mechanical adjustment of a small compensation capacitor C_P, which is connected in parallel with a 9 MΩ resistor. The outer conducting sheath of the cable has an earth clip connected to it, usually close to the point where the cable enters the probe-housing.

The probe is correctly adjusted when v_2 is a scaled replica of v_1. For this to occur it is necessary that $C_P R_P = (C_T + C_I)R_I$. This condition is achieved, in practice, by observing the CRO display, when the probe is used to monitor a

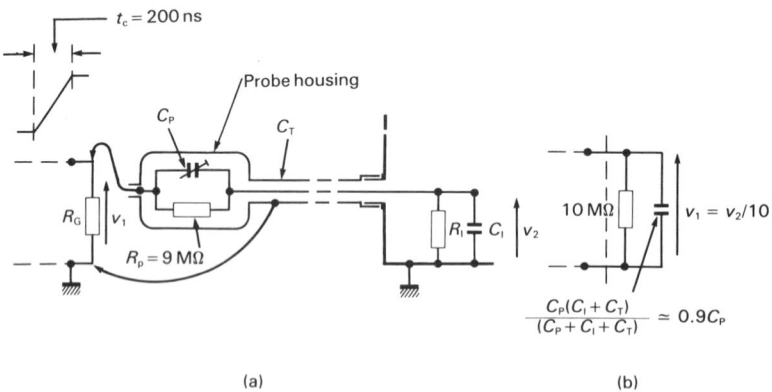

(a)

(b)

Fig. C.2 (a) Use of 10 × probe (signal attenuated by factor of 10); (b) Load across R_G for correct probe adjustment, $C_p R_p = (C_T + C_I)R_I$.

square-wave, voltage-calibration signal normally available at a front panel connection of the CRO, and making a mechanical adjustment so that there is no overshoot or undershoot on the edges of the observed waveform. In one form of mechanical adjustment a screwdriver (preferably non-metallic) is used to set the position of a screw, and thus effectively alter the dimensional separation of the plates of the small trimmer capacitor C_P, via a hole in the probe-housing provided for that purpose.

After the probe has been adjusted correctly the attenuation factor A $(= v_1/v_2)$ = 10 and the equivalent load presented across R_G is 10 MΩ in parallel with a capacitance slightly less than C_P, as shown in Fig. C.2(b) for the normal case $(C_T + C_l) \gg C_P$: typically, $C_P \approx$ a few pF.

Active probes offering a high input resistance and low input capacitance without significant attenuation are available as CRO accessories from some manufacturers (e.g. Hewlett-Packard). They can also be constructed in the laboratory using field-effect and bipolar junction transistors, but they generally require auxiliary power supplies which may not be available from the circuit under test.

DISTRIBUTED SYSTEMS

Suppose, in Fig. C.1, that t_c = 2 ns instead of 200 ns as considered above. The condition $2t_d \ll t_c$ no longer holds, the system must obviously be regarded as distributed, and the set-up of Fig. C.1(a), without modification, would be unsuitable for waveform monitoring. There would generally be reflections at the ends of the oscilloscope leads which could interfere with, and possibly cause malfunction of, a circuit under test – particularly a logic gate. The waveform observed will have transition times limited, of course, by the oscilloscope rise-time. The arrangement of Fig. C.2(a) is also *unsatisfactory* because the condition $C_P R_P = (C_T + C_l)R_l$ depends on the assumption of a lumped system which is not true in this case.

To minimize, if not eliminate completely, reflections at a CRO input terminal in standard run-of-the-mill laboratory CROs, real-time wide-band CROs and sampling CROs, it is obligatory to use correctly terminated CRO leads. For a low-cost CRO a standard, commercially available, 50 Ω termination can be connected via a BNC 'T-piece' as shown in Fig. C.3(a).

For a sampling oscilloscope, with which we will be concerned for the remainder of this appendix, the 50 Ω is a built-in feature. Input connections can be of the BNC or the 'hermaphrodite' General Radio (GR) variety. There are no real problems here since GR/BNC (male) and GR/BNC (female) connections are readily available. For simplicity in drawing we will show, schematically in section form, BNC connectors only.

However, the arrangement in Fig. C.3(b) is still not suitable as it stands

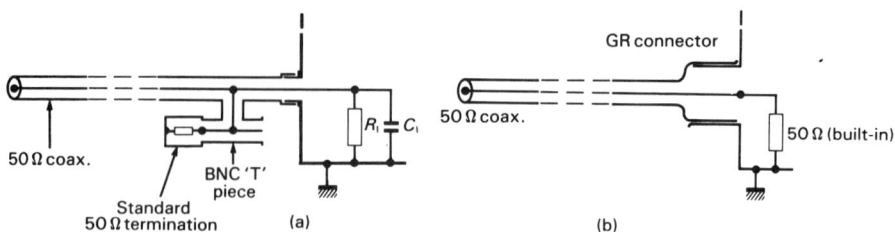

Fig. C.3   Oscilloscope connections for fast-edge monitoring. (a) Basic laboratory CRO; (b) High-speed CRO, e.g. Sampling CRO. A GR input connector is often used but BNC→GR and GR→BNC adaptors are available.

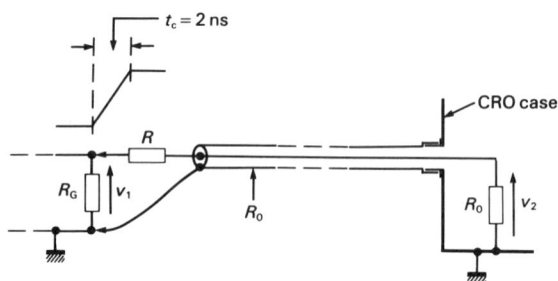

Fig. C.4   Passive probe for observing waveform across R_G. $(R+R_0) \ll R_G$.

because it presents to the test circuit, at the input to the cable, an impedance of 50 Ω which will cause problems if it is required to monitor the waveform across a resistor, R_G, having a value comparable with 50 Ω. Clearly, a series resistor, R, must be incorporated to give, with the cable input impedance, R_0, a potential divider that gives a known, convenient, attenuation factor and does not load, significantly, the circuit under test (see Fig. C.4). R_G is not heavily loaded if R is chosen so that $(R + R_0) \gg R_G$, i.e. $(R + R_0) \geqslant 10R_G$. This is the principle of the wide-band passive probe.

Table C.1 lists some useful values of R and associated attenuation factor A = (v_1/v_2) = $\{(R + R_0)/R_0\}$, for the case $R_G = 50$ Ω, $R_0 = 50$ Ω. For $R_G = 50$ Ω

Table C.1
Convenient values of R and A for passive probe of Fig. C.4.

R (Ω)	$R + R_0$ (Ω)	$A = (v_1/v_2)$
1200	1250	25
950	1000	20
450	500	10

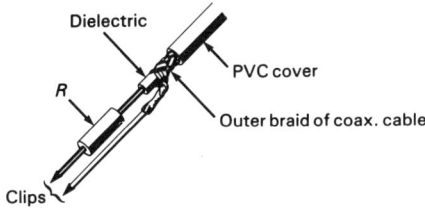

Fig. C.5   A practical realization of probe tip in Fig. C.4.

and $R = 450 \, \Omega$ the loading effect of the probe is such as to give a voltage reflection coefficient of about -0.05 for a 50 Ω test circuit cable that R_G is intended to match. This is often acceptable but lower values of R are generally not advisable.

A passive probe is a purchasable item. However, the reader may wish to construct a simple low-cost version, as in Fig. C.5, for himself. A low-inductance resistor, R is soldered to the inner conductor of the coaxial cable, and the outer braid is twisted to form a lead that is soldered to an earth clip, 'prod', or hook. All wires must, of course, be kept as short as possible. In a more elaborate form of the probe, R can be surrounded by a cylindrical metal sheath soldered to the cable braid. In this design the sheath can surround all but the probe-tip end of R and be separated from it by an insulating sleeve. An earth clip is soldered to the metal cylinder. For a well-made probe the signal displayed on the CRO screen is merely an attenuated and delayed version of the actual circuit waveform.

It is important to remember that the set-up of Fig. C.4 is not suitable if the signal being monitored does not have an earth rest level. Such is the case with the signal from an ECL gate. Direct connection of the CRO passive probe, just described, might upset the d.c. conditions in the circuit under test. However, this difficulty is overcome by using a series coupling capacitor, shown as C_c in Fig. C.6. The magnitude of C_c depends on the maximum pulse width, t_p, that it is required to display without appreciable 'sag' or 'droop' on the waveform. To avoid significant sag it is necessary that $C_c(R + R_0) \gg t_p$. Typically, C_c is in the range 1 to 100 nF.

It may be possible, in certain circumstances, to obtain a waveform for observation without any loading at all on the circuit under test. This is the

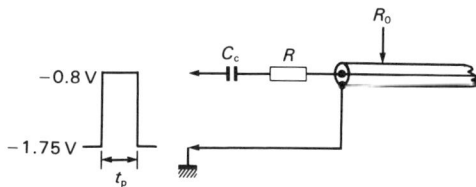

Fig. C.6   Use of coupling capacitor, C_c, to observe fast signals with significant d.c. bias levels.

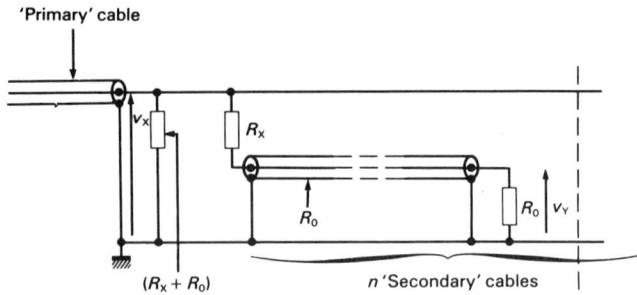

Fig. C.7 A 'signal-splitter' circuit can be used for waveform observation: $R_X = nR_0$; $(v_X/v_Y) = n + 1$.

case for a signal-splitting technique exemplified in Fig. C.7. The 'Primary' cable in the test circuit and each of the n 'Secondary' cables are all assumed to have the same characteristic resistance R_0. One or more of the secondary cables can be used for waveform monitoring. It is easily shown that all of the cables are correctly terminated if $R_X = nR_0$: thus, if $R_0 = 50\ \Omega$ and $n = 2$, then $R_X = 100\ \Omega$ and $A = (v_X/v_Y) = 3$.

Occasionally it is necessary to employ an attenuator to reduce the magnitude of a fast pulse signal which it is required to observe. This is the case when the pulse source is the output from a mercury-wetted relay pulse generator and an 'avalanche' pulse generator. Connection, even via a 10 X passive probe, to the input circuit of a sampling oscilloscope could well destroy the input stage of the instrument since it is normally capable of accepting a signal in the range of only approximately ± 2 V about earth.

Switched attenuator boxes are available but fixed attenuator pads are easily constructed. These can be of the 'T' or 'π' variety. A symmetrical-T type attenuator is shown in Fig. C.8. This is required to operate between two sections of cable each with characteristic resistance R_0. It can be shown that Z_{ab}, the impedance seen looking to the right between input terminals a and b, is equal to R_0, the load across the output terminals c and d, if R_1 and R_2 are chosen so that the following relationship holds.

$$\sqrt{\{R_1 (R_1 + 2R_2)\}} = R_0 \tag{C.1}$$

Then the attenuation factor $A = (v_1/v_2)$ is given by,

$$A = (R_0 + R_1 + R_2)/R_2 \tag{C.2}$$

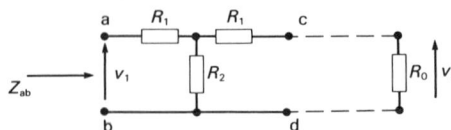

Fig. C.8 Symmetrical-T attenuator pad.

Equations (C.1) and (C.2) are useful for *checking* a given design, such as might be specified on a data sheet. But what if it is required to *design* an attenuator for a given R_0 and A?

The relevant equations are,

$$R_1 = R_0(A - 1)/(A + 1) \tag{C.3}$$

$$\text{and, } R_2 = 2R_0A/(A^2 - 1) \tag{C.4}$$

$$\text{where, as before, } A = (v_1/v_2) \tag{C.5}$$

In any problem involving attenuator pad design the requirements will generally be met only by using non-preferred resistor values. The choice then is to use selected, measured, components chosen from large batches or to choose close-tolerance nearest-preferred values and sacrifice some accuracy. The course selected depends on the problem.

Thus, if we require $A = 2$ when $R_0 = 50$ Ω, a calculation gives $R_1 = 16.7$ Ω, $R_2 = 66.7$ Ω. These are not preferred in any range. Suppose, however, we choose $R_1 = 16$ Ω and $R_2 = 68$ Ω. From eqn (C.1), $\sqrt{\{R_1 (R_1 + 2R_2)\}} = 49.3$ Ω, and from eqn (C.2), $A = 1.96$ which is close to $A = 2$. Two other useful sets of data using preferred values based on $R_0 = 50$ Ω are: $A \approx 5$ for $R_1 = 33$ Ω and $R_2 = 20$ Ω; $A \approx 10$ for $R_1 = 39$ Ω and $R_2 = 10$ Ω. 'T' stages can be cascaded, in which case their attenuation factors are multiplied.

An alternative to the symmetrical-T is the symmetrical-π pad of Fig. C.9. This may give more convenient resistor values in a particular application.

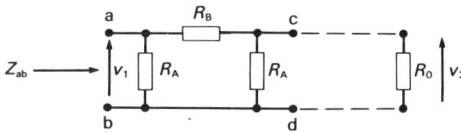

Fig. C.9   Symmetrical-π attenuator pad.

It can be shown that $Z_{ab} = R_0$ if,

$$R_A/\sqrt{\{1 + (2R_A/R_B)\}} = R_0 \tag{C.6}$$

in which case,

$$A = (v_1/v_2) = \{1 + (R_A/R_0) + (R_B/R_A)\} \tag{C.7}$$

Alternatively, we can write,

$$R_A = R_0(A + 1)/(A - 1) \tag{C.8}$$

$$\text{and, } R_B = R_0(A^2 - 1)/2A \tag{C.9}$$

Answers (including worked solutions to problems)

CHAPTER 1

P1.1 From the data given, $t_d = l/u = 3/(3 \times 10^8)$ S $= 10$ ns
Thus, $2t_d = 20$ ns.
    a conventional circuit theory is applicable for $t_c \gg 2t_d$. A safe choice is $t_c = 10(2t_d)$, i.e. $t_c \geqslant 0.2 \ \mu s$.
    b conventional circuit theory is definitely not applicable if $t_c \ll 2t_d$. A safe choice is $t_c = (2t_d/10) = 2$ ns.

P1.2 $t_c = 2lt_u$ when $l = t_c/2t_u$
Now $t_u = 5$ ns/m ≈ 5 ns/39 in
$\therefore \ l \approx (1.5 \text{ ns} \times 39 \text{ in})/(2 \times 5)$ ns ≈ 6 inches.

P1.3 In Fig. S1.3, straight line 0C corresponds to $2t_d = t_c$; horizontally shaded region gives $2t_d \geqslant 10t_c$; vertically shaded region gives $2t_d \leqslant (t_c/10)$.

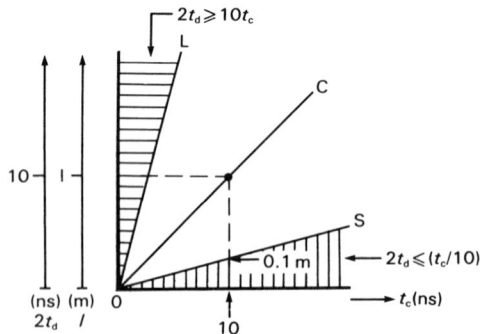

Fig. S1.3 $2t_d = 2lt_u$; for $t_u = 5$ ns/m, $2t_d = 10$ ns when $l = 1$ m.

CHAPTER 2

P2.1 $[L] = [V] [T]/[I]$
$[C] = [I] [T]/[V]$
Hence,
a $[L/C] = [V] [V]/[I][I]$
   and $[\sqrt{(L/C)}] = [V]/[I] = [R]$
Similarly,
b $[\sqrt{(LC)}] = [T]/$unit-length, because L, C are per-unit-length.

P2.2 **a** $Z_0 = \sqrt{(L/C)}$ as the line is lossless
$$Z_0(\Omega) = \sqrt{\left(\frac{73.75 \times 10^{-9}}{29.5 \times 10^{-12}}\right)} = \sqrt{\left(\frac{73750}{29.5}\right)} = 50\ \Omega$$
b $t_u = \sqrt{(LC)} = \sqrt{(73.75 \times 10^{-9} \times 29.5 \times 10^{-12})}$ s/ft
$= (10^{-9}) \sqrt{(0.7375 \times 2.95)}$ s/ft
$= 1.475$ ns/ft
c $t_u = 1.475 \times 3.281$, as 1 m ≈ 3.281 ft
≈ 4.84 ns/m

P2.3 $Z_0 = \sqrt{(L/C)}$ and $t_u = \sqrt{(LC)} = \sqrt{C}(Z_0\sqrt{C}) = Z_0 C = 72\ \Omega \times$
69 pF/m

$\therefore t_u \approx 5$ ns/m

P2.4 **a** $v = f\{t - x\sqrt{(LC)}\} = f(y)$, say, where $y = \{t - x\sqrt{(LC)}\}$
$$\therefore \left(\frac{\partial v}{\partial x}\right) = \left(\frac{dv}{dy}\right)\left(\frac{\partial y}{\partial x}\right)$$
$$= f'(y)\{-\sqrt{(LC)}\}, \text{ where } f'(y) = \frac{df(y)}{dy}$$

Extending this argument,

$$\left(\frac{\partial^2 v}{\partial x^2}\right) = f''(y)\{\sqrt{(LC)}\}\{\sqrt{(LC)}\} = f''(y)\{LC\}$$

where $f''(y) = \dfrac{d^2f(y)}{dy^2}$

Also, $\left(\dfrac{\partial v}{\partial t}\right) = \left(\dfrac{dv}{dy}\right)\left(\dfrac{\partial y}{\partial t}\right) = f'(y)$, as $\left(\dfrac{\partial y}{\partial t}\right) = +1$

and $\left(\dfrac{\partial^2 v}{\partial t^2}\right) = f''(y)$

Thus, $\left(\dfrac{\partial^2 v}{\partial x^2}\right) = LC\left(\dfrac{\partial^2 v}{\partial t^2}\right)$

or, $\dfrac{1}{LC}\left(\dfrac{\partial^2 v}{\partial x^2}\right) = \left(\dfrac{\partial^2 v}{\partial t^2}\right)$ which is eqn (2.8) in the text.

b and **c** follow by a similar argument, i.e. putting $y = t + x\sqrt{(LC)}$ in **b** and $y = \{t + x\sqrt{(LC)} + X\sqrt{(LC)}\}$ in **c**

d Let $v = f(y) + g(w)$, where $y = t - x\sqrt{(LC)}$ and $w = t + x\sqrt{(LC)}$

$$\therefore \left(\frac{\partial v}{\partial x}\right) = f'(y)\{-\sqrt{(LC)}\} + g'(w)\{-\sqrt{(LC)}\}$$
$$= \{-\sqrt{(LC)}\}\{f'(y) + g'(w)\}$$

Similarly,

$$\left(\frac{\partial^2 v}{\partial x^2}\right) = LC\{f''(y) + g''(w)\}$$

Also,

$$\left(\frac{\partial v}{\partial t}\right) = \{f'(y) + g'(w)\} \text{ as } \left(\frac{\partial y}{\partial t}\right) = \left(\frac{\partial g}{\partial t}\right) = +1$$

and $\left(\dfrac{\partial^2 v}{\partial t^2}\right) = \{f''(y) + g''(w)\}$

Thus, $\left(\dfrac{\partial^2 v}{\partial x^2}\right) = LC\left(\dfrac{\partial^2 v}{\partial t^2}\right)$

or, $\dfrac{1}{LC}\left(\dfrac{\partial^2 v}{\partial x^2}\right) = \left(\dfrac{\partial^2 v}{\partial t^2}\right)$ which, again, is eqn (2.8)

P2.5 **a** When the step is at a point x' on the line the energy stored in the electric field is equal to (Total capacitance to point x') $V^2/2$.

$$\therefore W_E = (Cx')V^2/2$$

b The energy stored in the magnetic field is equal to (Total inductance to point x') $I^2/2$

$$\therefore W_M = (Lx')I^2/2$$

c $W_E = (Cx')V^2/2 = \left(\dfrac{L}{R_0^2}\right)x'\left(\dfrac{IR_0}{2}\right)^2$ as $\sqrt{C} = \dfrac{\sqrt{L}}{R_0}$

and $V = I_0 R$.

$$\therefore W_E = (Lx')I^2/2 = W_M$$

d Total energy supplied to line when voltage step has reached point x' along it is W.

$$W = W_E + W_M = 2W_E, \text{ from P2.5 c, above.}$$

Now, the rate at which energy is increasing is the power, P, continuously supplied to the line, as the wavefront moves from x' to $(x' + \delta x')$ in a time δt.

$$P = \left(\frac{dW}{dt}\right) = \left(\frac{2dW_E}{dt}\right) = \left(\frac{2dW_E}{dx'}\right)\left(\frac{dx'}{dt}\right) = CV^2u$$

But, $u = 1/\sqrt{(LC)}$ and $R_0 = \sqrt{\dfrac{L}{C}}$

$$\therefore P = CV^2/\sqrt{(LC)} = V^2\sqrt{\frac{C}{L}} = V^2/R_0 = VI$$

P2.6 By analogy with the argument leading to eqn (2.1),

$$v - (v + \delta v) = \{(L/2)\delta x + (L/2)\delta x\}(\delta i/\delta t) + \{(R/2)\delta x + (R/2)\delta x\}i$$
$$\therefore -\delta v = \{L(\delta x)(\delta i/\delta t)\} + R(\delta x)i$$

In the limit,

$$-\left(\frac{\partial v}{\partial x}\right) = Ri + L\left(\frac{\partial i}{\partial t}\right)$$

Similarly,

$$i - (i + \delta i) = \{(C\delta x)(\delta v/\delta t)\} + (G\delta x)v$$
$$\therefore -\delta i = \{C(\delta x)(\delta v/\delta t)\} + (G\delta x)v$$

In the limit,

$$-\left(\frac{\partial i}{\partial x}\right) = Gv + C\left(\frac{\partial v}{\partial t}\right)$$

P2.7 From problem (P2.6),

$$\left(\frac{\partial^2 v}{\partial x^2}\right) = -R\left(\frac{\partial i}{\partial x}\right) - L\frac{\partial}{\partial x}\left(\frac{\partial i}{\partial t}\right) \tag{P2.7.1}$$

Also, $$-\frac{\partial}{\partial t}\left(\frac{\partial i}{\partial x}\right) = G\left(\frac{\partial v}{\partial t}\right) + C\left(\frac{\partial^2 v}{\partial t^2}\right) \tag{P2.7.2}$$

$$\therefore -L\frac{\partial}{\partial t}\left(\frac{\partial i}{\partial x}\right) = LG\left(\frac{\partial v}{\partial t}\right) + LC\left(\frac{\partial^2 v}{\partial t^2}\right) \tag{P2.7.3}$$

From eqns (P2.7.1) and (P2.7.3)

$$\left(\frac{\partial^2 v}{\partial x^2}\right) = -R\left(\frac{\partial i}{\partial x}\right) + LG\left(\frac{\partial v}{\partial t}\right) + LC\left(\frac{\partial^2 v}{\partial t^2}\right) \tag{P2.7.4}$$

But, $$-\left(\frac{\partial i}{\partial x}\right) = Gv + C\left(\frac{\partial v}{\partial t}\right)$$

$$\therefore -R\left(\frac{\partial i}{\partial x}\right) = RGv + RC\left(\frac{\partial v}{\partial t}\right) \tag{P2.7.5}$$

Hence, from eqns (P2.7.4) and (P2.7.5), after some re-arrangement of terms,

$$\left(\frac{\partial^2 v}{\partial x^2}\right) = LC\left(\frac{\partial^2 v}{\partial t^2}\right) + (RC + LG)\left(\frac{\partial v}{\partial t}\right) + RGv$$

P2.8 Assume $v = \exp(-\alpha x)f\left(t - \frac{x}{u}\right)$ (P2.8.1)

where, $\alpha = R\sqrt{\dfrac{C}{L}}$ and $u = 1/\sqrt{(LC)}$

then,

$$\left(\frac{\partial v}{\partial x}\right) = -\alpha \exp(-\alpha x)f\left(t - \frac{x}{u}\right) - \frac{1}{u}\exp(-\alpha x)f'\left(t - \frac{x}{u}\right)$$

$$= \exp(-\alpha x)\left\{-\alpha f\left(t - \frac{x}{u}\right) - \frac{1}{u}f'\left(t - \frac{x}{u}\right)\right\} \quad \text{(P2.8.2)}$$

$$\left(\frac{\partial^2 v}{\partial x^2}\right) = -\alpha\exp(-\alpha x)\left\{-\alpha f\left(t - \frac{x}{u}\right) - \frac{1}{u}f'\left(t - \frac{x}{u}\right)\right\}$$

$$+ \exp(-\alpha x)\left\{\frac{\alpha}{u}f'\left(t - \frac{x}{u}\right) + \frac{1}{u^2}f''\left(t - \frac{x}{u}\right)\right\} \quad \text{(P2.8.3)}$$

Also, $\left(\dfrac{\partial v}{\partial t}\right) = \exp(-\alpha x)f'\left(t - \dfrac{x}{u}\right)$ (P2.8.4)

$$\left(\frac{\partial^2 v}{\partial t^2}\right) = \exp(-\alpha x)f''\left(t - \frac{x}{u}\right) \quad \text{(P2.8.5)}$$

For the case $RC = GL$ the final expression in problem P2.7 reduces to

$$\left(\frac{\partial^2 v}{\partial x^2}\right) = LC\left(\frac{\partial^2 v}{\partial t^2}\right) + 2RC\left(\frac{\partial v}{\partial t}\right) + RGv \quad \text{(P2.8.6)}$$

Using eqns (P2.8.2) to (P2.8.5) we have to show that eqn (P2.8.1) satisfies eqn (P2.8.6).

Now eqn (P2.8.3) can be rewritten,

$$\left(\frac{\partial^2 v}{\partial x^2}\right) = \exp(-\alpha x)\left[\alpha^2 f\left(t - \frac{x}{u}\right) + \frac{2\alpha}{u}f'\left(t - \frac{x}{u}\right)\right.$$

$$\left. + \frac{1}{u^2}f''\left(t - \frac{x}{u}\right)\right] \quad \text{(P2.8.7)}$$

$$= \alpha^2 v + \frac{2\alpha}{u}\left(\frac{\partial v}{\partial t}\right) + \frac{1}{u^2}\left(\frac{\partial^2 v}{\partial t^2}\right) \quad \text{(P2.8.8)}$$

Eqn (P2.8.8) makes use of eqns (P.2.8.4) and (P2.8.5)

Now $\alpha^2 = R^2C/L = RG$; $2\alpha/u = 2R\sqrt{\dfrac{C}{L}} \cdot \sqrt{(LC)} = 2RC$; $\dfrac{1}{u^2} = LC$

$$\therefore \left(\frac{\partial^2 v}{\partial x^2}\right) = LC\left(\frac{\partial^2 v}{\partial t^2}\right) + 2RC\left(\frac{\partial v}{\partial t}\right) + RGv, \text{ which is eqn}$$

$$\text{(P2.8.6)}$$

CHAPTER 3

P3.1   The 'boundary conditions' at $x = 0$ and $x = l$ give:

$\rho_{VG} = 0; \rho_{VT} = -1$

Hence from eqns (2.33), (2.37) of Chapter 2,

$\rho_{IG} = 0; \rho_{IT} = +1$

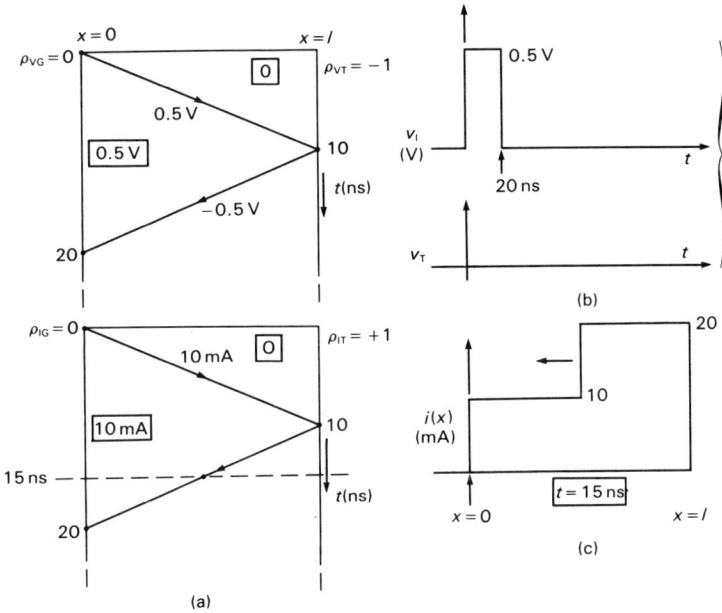

Fig. S3.1.

Figure S3.1(a), (b), (c) illustrate the solutions to parts (a), (b), (c), respectively, of the problem.

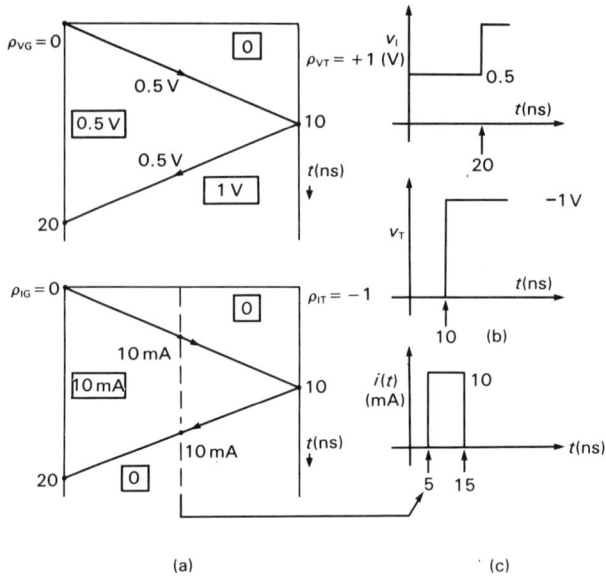

Fig. S3.2.

P3.2   See Fig. S3.2.

P3.3   See Fig. S3.3.

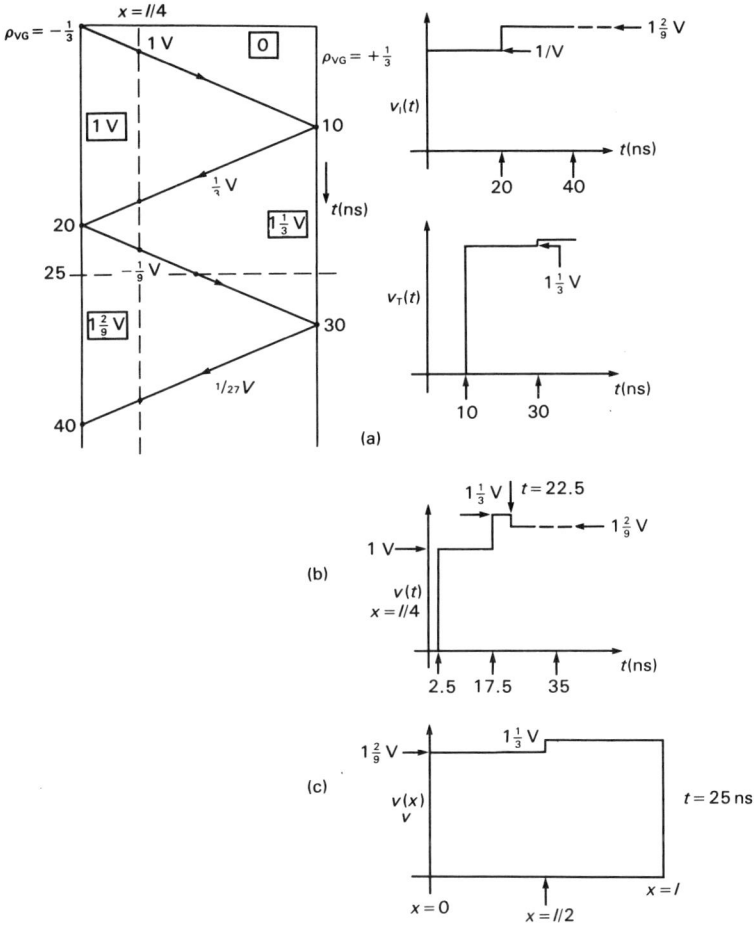

Fig. S3.3.

P3.4   The reflection chart is shown in Fig. S3.4(a). At $t = 10$ ns, Sw opens so the net input current is zero. However, the existing current is 20/3 mA. Hence a forward current of $-20/3$ mA, corresponding to a forward voltage of $-1/3$ V, must traverse the line. Waveforms are shown in Fig. S3.4(b).

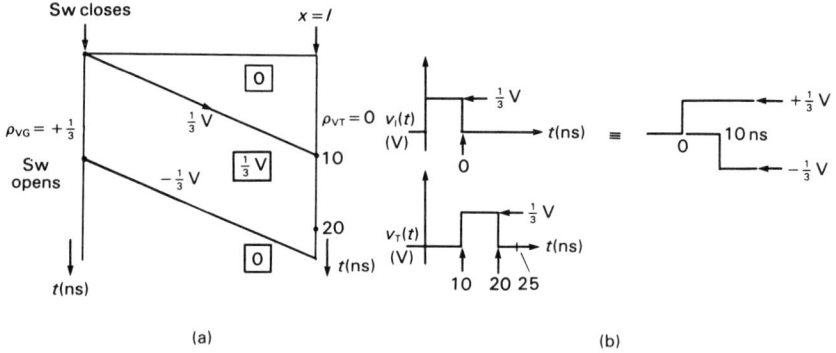

(a)   (b)

Fig. S3.4

P3.5   See Fig. S3.5.

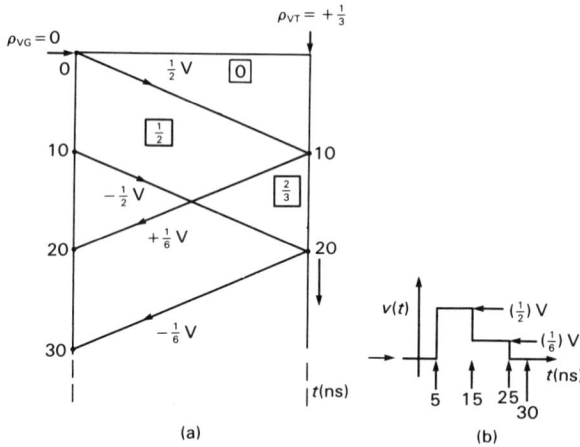

(a)   (b)

Fig. S3.5

P3.6   Application of the Thévenin–Norton transformation in Fig. S3.6(a)
gives v_G in (b). This, as shown in (c), can be resolved into positive-
and negative-going step-edges. The reflection chart is shown in (d)
and from this we obtain the required $v_l(t)$ shown in (e). For
40 ns $> t > 20$ ns the line voltage is zero and the energy stored in the
electric field is also zero. However, energy is stored in the magnetic
field and is dissipated in the source resistance for the period
60 ns $\geqslant t \geqslant 40$ ns.

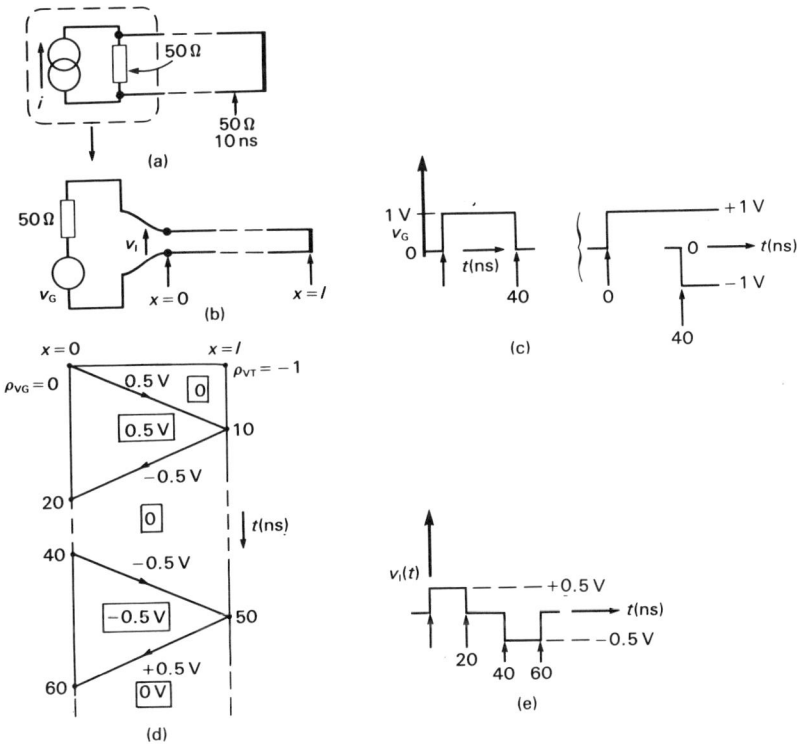

Fig. S3.6

P3.7   Referring to Fig. 3.6 of the text:
For 40 ns $> t >$ 20 ns the line current is 40 mA. Opening Sw imposes
the boundary condition that the net line current at $x = 0$ is zero. This
is achieved if a -40 mA current step starts down the line. On a
voltage reflection chart, which we plot because we are interested in
$v_1(t)$, this corresponds to a voltage step $= -(40$ mA $\times 50 \ \Omega) = -2$ V.

The voltage reflection chart is shown in Fig. S3.7(a). For completeness, behaviour is considered from $t = 0$. Figs S3.7(b), S3.7(c)
show $v_1(t)$, $v_T(t)$, respectively.

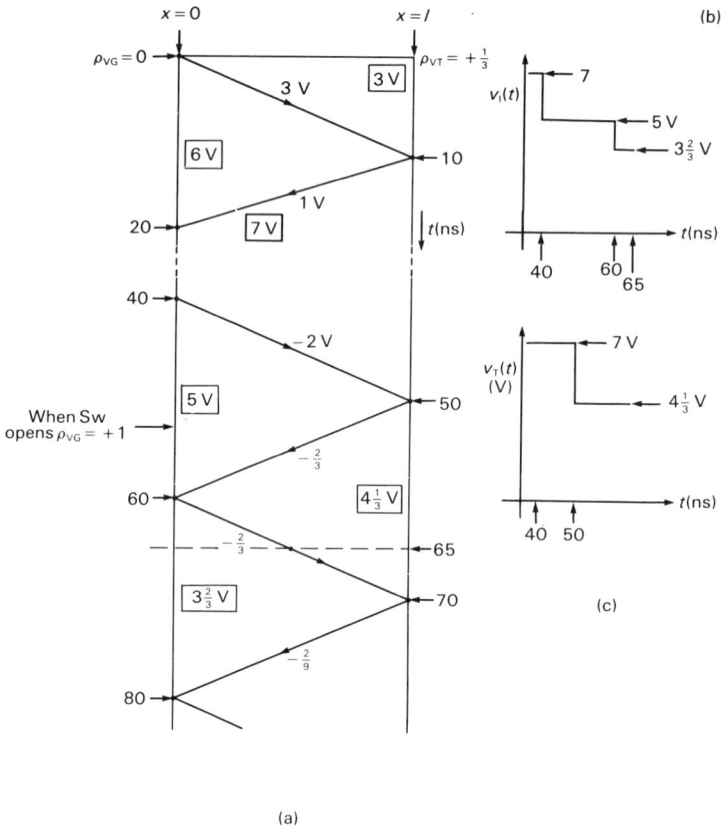

(a)

(b)

(c)

Fig. S3.7

P3.8 The voltage change Δv_I is the sum of incident and reflected waveforms.

In Fig. S3.8: $\alpha = \{R_0/(R_0 + R_G)\}$; $\rho_{VG} = \{(R_G - R_0)/(R_G + R_0)\}$

a For $r = 1, 2, 3, \ldots n.$

$$\Delta v_I(2t_d) = \alpha V(1 + \rho_{VG}) = \{R_0/(R_0 + R_G)\}\{1 + \rho_{VG}\}\ V$$

b $\quad \Delta v_I\{2(2t_d)\} = \alpha V(1 + \rho_{VG})\rho_{VG}$

$\quad\quad \Delta v_I\{3(2t_d)\} = \alpha V(1 + \rho_{VG})\rho_{VG}^2$

$$\vdots$$

$$\Delta v_I\{r(2t_d)\} = \alpha V(1 + \rho_{VG})\rho_{VG}^{r-1}$$

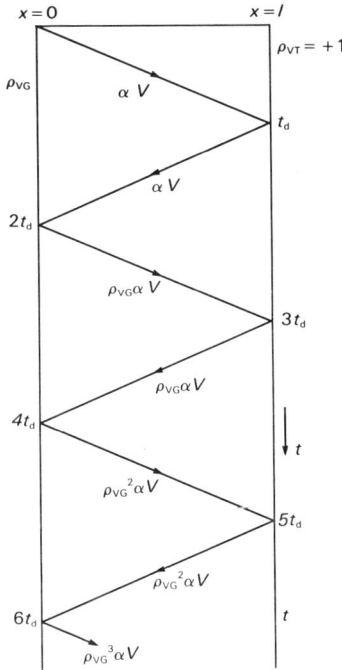

Fig. S3.8

It follows that,

$$[\Delta v_I\{r(2t_d)\}]\ [\Delta v_I\{(r-1)\,(2t_d)\}] = \rho_{VG}$$

P3.9   For $R_G \gg R_0$,

$$v_I(t) = \sum_{r=1}^{r-n} \Delta v_I\{r(2t_d)\}$$

$$v_I(t) = \alpha V(1 + \rho_{VG})\{1 + \rho_{VG} + \rho_{VG}^2 \ldots \ldots \rho_{VG}^{n-1}\}$$

and, $\rho_{VG}v_I(t) = \alpha V(1 + \rho_{VG})\{\rho_{VG} + \rho_{VG}^2 \ldots \ldots \rho_{VG}^n\}$

$$\therefore v_I(t)\{1 - \rho_{VG}\} = \alpha V(1 + \rho_{VG})\{1 - \rho_{VG}^n\}$$

$$\therefore v_I(t) = v_I\{n(2t_d)\} = \alpha V(1 + \rho_{VG})\{1 - \rho_{VG}^n\}/(1 + \rho_{VG})$$

Now $\alpha(1 + \rho_{VG})/(1 - \rho_{VG}) = \{R_0/(R_0 + R_G)\}\{R_G/R_0\} = R_G/(R_G + R_0)$
$$\approx 1, \text{ as } R_G \gg R_0.$$

$$\therefore v_I(t) \approx V(1 - \rho_{VG}^n)$$

P3.10   **a** $\rho_{VG}^n = \exp(-t/\tau)$, given.

$$\therefore n \log_e \rho_{VG} = -t/\tau$$

But $n \log_e \left[\dfrac{R_G - R_0}{R_G + R_0}\right] = n[\log_e\{1 - (R_0/R_G)\} - \log_e\{1 + (R_0/R_G)\}]$

$$\approx n[-(R_0/R_G) - (R_0/R_G) + \text{negligible higher-order terms}]$$

$$\approx -2nR_0/R_G$$

$$\therefore \frac{2nR_0}{R_G} \approx \frac{t}{\tau}$$

However, $t = 2nt_d$

$$\therefore \frac{2nR_0}{R_G} = \frac{2nt_d}{\tau}$$

$$\therefore \tau = t_d(R_G/R_0)$$

If C_T is the total line capacitance, $t_d = (C_T/C)\sqrt{(LC)}$
Substituting this and $R_0 = \sqrt{(L/C)}$ in the expression for τ,

$$\tau \approx R_G C_T$$

b It follows from problem P3.9 and part **a**, above, that

$$v_I(t) \approx V[1 - \exp(-t/C_T R_G)]$$

The interpretation of this result is that, for an open-circuited line and $R_G \gg R_0$, $t_d \ll C_T R_G$ and the incremental voltage steps at the

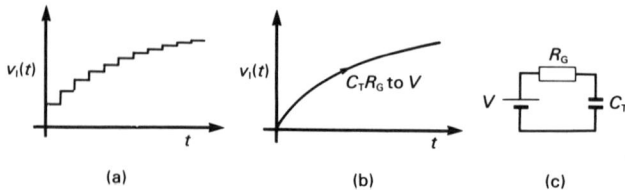

Fig. S3.10

input are very small in comparison with V. The charging of the line, actually of the form indicated in Fig. S3.10(a), approaches the form of Fig. S3.10(b), corresponding to an effective equivalent circuit shown in Fig. S3.10(c).

P3.11   See Fig. S3.11(a) to (d).

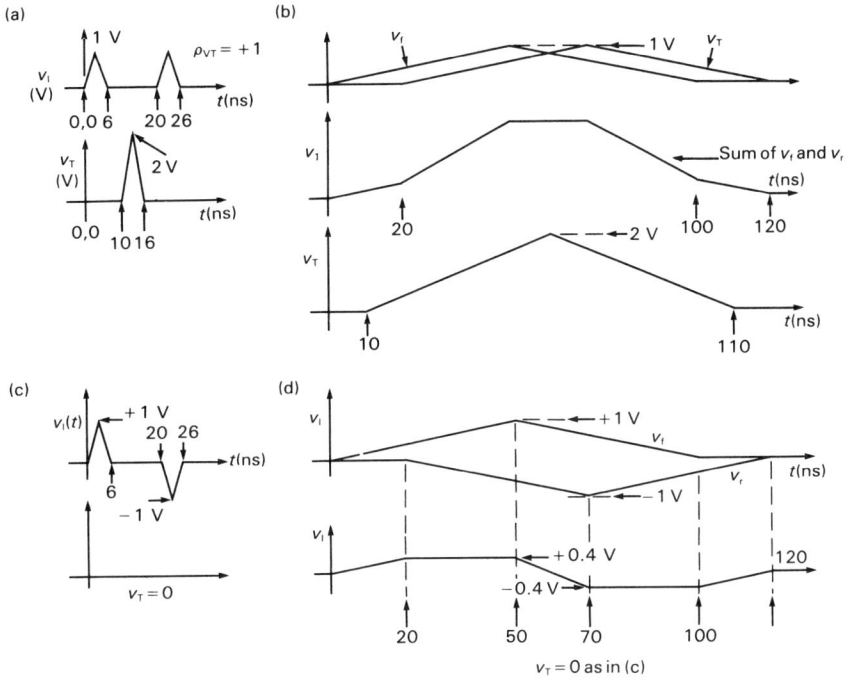

Fig. S3.11

CHAPTER 4

P4.1   Figure S4.1(a) gives the graphical construction for finding $v_I(t)$, $v_0(t)$; Fig. S4.1(b) shows the waveforms obtained; Fig. S4.1(c) is the reflection chart.

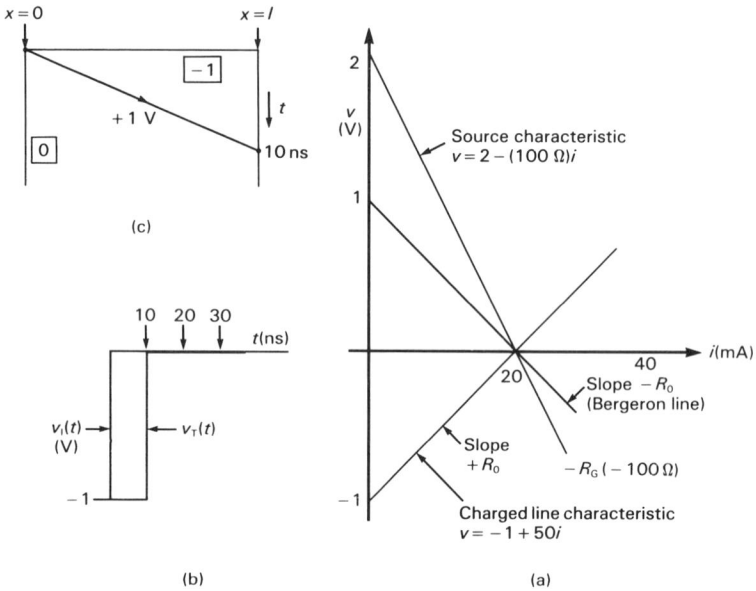

Fig. S4.1

P4.2   It is instructive to regard v_Y as the *input* voltage to the line. Fig. S4.2(a) and (b) are solutions to parts **a** and **b** respectively, of the problem. The graphs are scaled so that the Bergeron lines are at right angles. If v_X is regarded as the input to the line — the usual convention — then the graphical solution is in the right half plane $i > 0$ (see Example 4.2 of text).

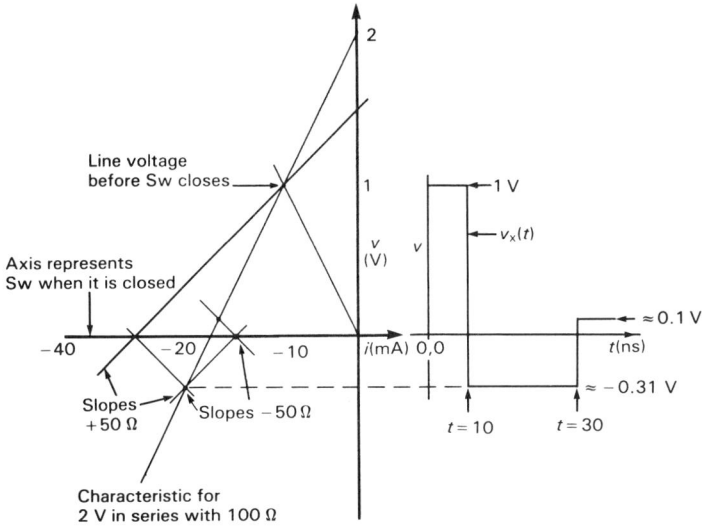

(a)

Fig. S4.2(a)

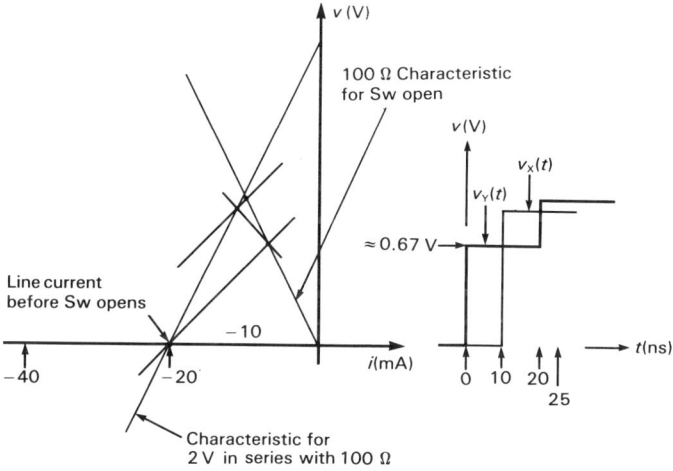

Fig. S4.2(b)

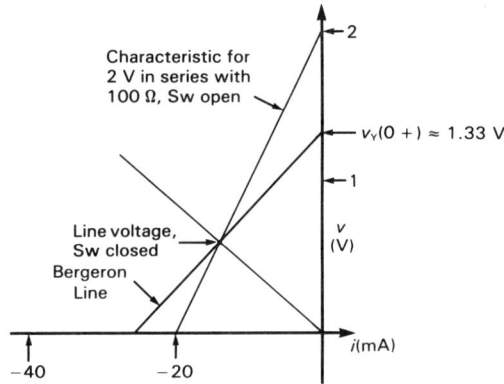

Fig. S4.3(a)

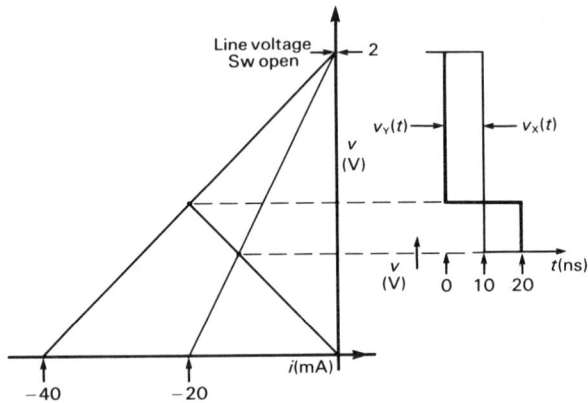

Fig. S4.3(b)

P4.3   Fig. S4.3(a) and (b) are solutions to **a** and **b** respectively. (See comment in problem P4.2, above, regarding the location of the plot in the v, i plane.)

P4.4   1 m = 3.281 ft. t_u = 1.475 ns/ft (given.)

$$\therefore t_d = 3.281 \times 1.475 = 4.84 \text{ ns.}$$

When Sw closes the initial circuit is shown in S4.4(a). A − 50 V step travels back along the line towards the 47 kΩ resistor. The reflection coefficient ρ_{VG} is

$$\rho_{VG} = \left[\frac{(47 \text{ k}\Omega - 50 \text{ }\Omega)}{(47 \text{ k}\Omega + 50 \text{ }\Omega)} \right] = [1 - (50 \text{ }\Omega/47 \text{ k}\Omega)]/\{1 + (50 \text{ }\Omega/47 \text{ k}\Omega)\}$$

$$\text{or, } \rho_{VG} \approx \left[1 - \left(\frac{100 \text{ }\Omega}{50 \text{ k}\Omega} \right) \right] = \left[1 - \left(\frac{2}{1000} \right) \right]$$

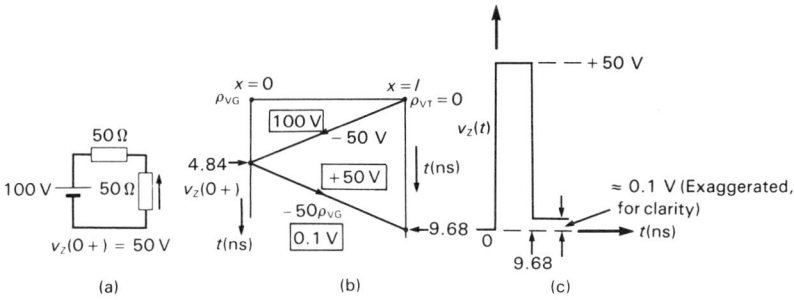

Fig. S4.4

There is no reflection at $t = 9.68$ ns because the line is correctly terminated at the switch end with Sw closed.

$$v_Z(9.68 \text{ ns}) \approx 50(1 - \rho_{VG}) \approx (50 \times 2/1000) \text{ V} = 0.1 \text{ V}$$

Figure S4.4(c) shows how the waveform $v_Z(t)$ is obtained from the Reflection Chart of Fig. S4.4(b).

P4.5   Figure S4.5 is self-explanatory,

$$v_Z(9.68 \text{ ns}) = (50 - 49.9 + 16.63) \text{ V} = 16.73 \text{ V}$$

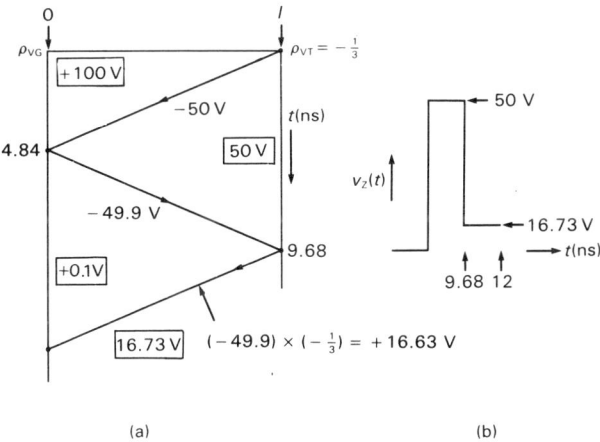

Fig. S4.5

P4.6   As the RG 8A/U is 1 m long it has a total line capacitance
$C_T = (29.5 \text{ pF/ft}) \times (3.281 \text{ ft}) \approx 100 \text{ pF}$

When Sw opens, $R_G = 47 \text{ k}\Omega \gg 50 \text{ }\Omega$, and the circuit for estimating $v_X(t)$ is shown in Fig. S4.6(a). The time constant $R_G C_T \approx 4.7 \text{ }\mu\text{s} \gg t_d$.

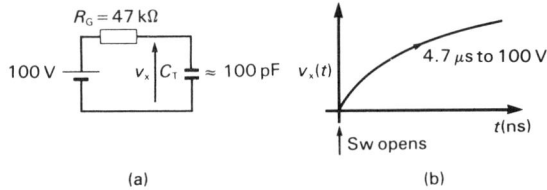

Fig. S4.6

As in the solution to problem P3.10 this means that the incremental voltage steps in v_X are minute in comparison with 100 V and the waveform for $v_X(t)$ resembles an exponential with time-constant 4.7 μs, as shown in Fig. S4.6(b). Obviously the line is not fully charged until an interval of $\approx 5 \times 4.7$ μs, say 25 μs, has passed after Sw opens. To obtain 50 V pulses across R_T on a repetitive basis means that Sw must operate at a pulse repetition frequency $< (1/25$ μs), i.e. 40 kHz.

P4.7   Refer to Fig. S4.7(a). 'a' the point of intersection of the output characteristic of gate A, in Fig. 4.18, and the input characteristic of gate B, gives the initial condition prior to the '0' to '1' transition.

The Bergeron lines describe the '0' to '1' transition: the resulting waveforms for $v_1(t)$, $v_T(t)$ are shown in Fig. S4.7(b).

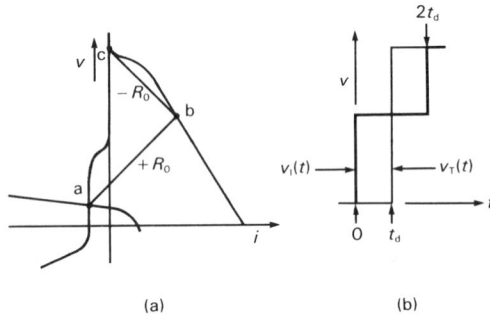

Fig. S4.7

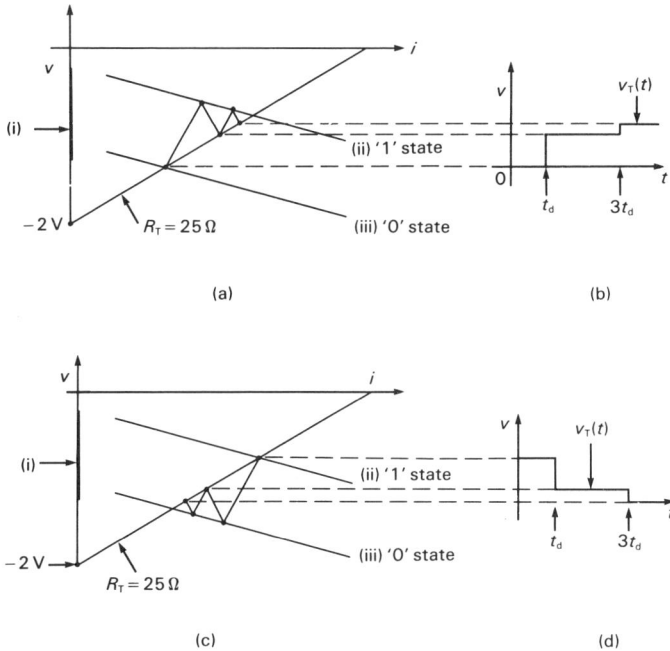

Fig. S4.8

P4.8   For the ECL gate set-up for Fig. 4.22(a) with $R_T = 25$ Ω, Fig. S4.8(a) shows the construction for a '0' to '1' transition. Bergeron lines correspond to slopes of $\pm R_0$. Fig. S4.8(b) for $v_T(t)$ is obtained from (a) by cross-projection. Similarly, Fig. S4.8(c), (d) show the constructions for $4t_d > t > 0 +$, for a '1' to '0' transition.

P4.9   At X the impedance seen looking to the right is (150 Ω ‖ 150 Ω) = 75 Ω. This matches the 75 Ω line to the left of X. Looking to the right at Y we see a correctly terminated 150 Ω line in parallel with a correctly terminated 75 Ω line. The impedance seen looking to the right at Y is thus (75 Ω‖150 Ω) = 50 Ω. This matches the 50 Ω line to the left of Y.

P4.10 Figure S4.10 explains, graphically, series matching using the SLL approach. In (a) the load R_p of Fig. 4.26 cuts the output characteristics in the '0' and '1' states at P and Q respectively. The addition of R_S (Fig. 4.26), if appropriately chosen, modifies the output characteristics from (ii) and (iii) to (ii)' and (iii)', respectively. The slopes of (ii)' and (iii)' are each, ideally, $- R_0$.

The arrows in (b) show the state of affairs for a '0' to '1' transition. The Bergeron lines are either coincident with or perpendicular to (ii)' and (iii)'. The resultant waveforms for a '0' to '1' transition are shown in (c) and those for a '1' to '0' transition in (d).

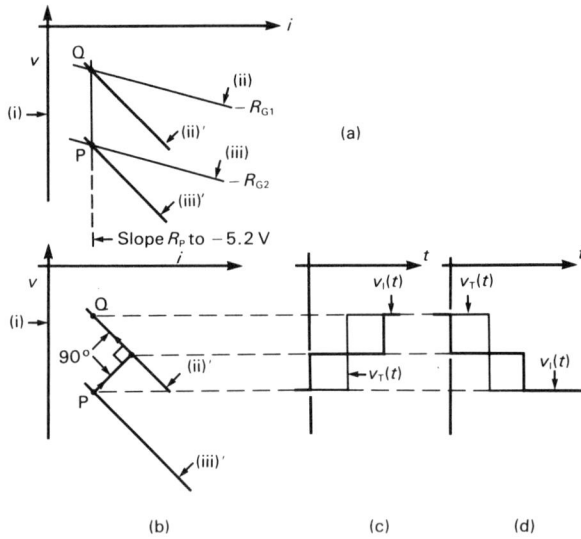

Fig. S4.10

CHAPTER 5

P5.1 Figure S5.1 shows $v_I(t)$: this is a redrawn version of Fig. 5.3(f). The exponential rise is given by:

$$v_I(t - 2t_d) = V[1 - \exp\{-(t - 2t_d)/R_0 C_L\}]$$

But, $t_k = (t - 2t_d)$

Thus, $v_I(t_k) = V[1 - \exp\{-t_k/R_0 C_L\}]$

$v_I(t_k) = V/2$ when

$$V/2 = V[1 - \exp\{-t_k/R_0 C_L\}]$$

Hence, $t_k = C_L R_0 \log_e 2$.

For $C_L = 100$ pF, $R_0 = 50\ \Omega$,

$t_k \approx 3.5$ ns

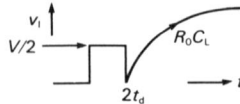

Fig. S5.1

P5.2   From eqn (2.11) of Chapter 2,

$v_f(x', t) = U(t - t')f(t - t')$, where $t' = (x'/u)$

From a Table of Laplace Transforms:

$v_f(x', s) = f(s) \exp(-st') = f(s)\exp(-sx'/u) = f(s)\exp(-sx't_u)$

P5.3 $\rho_{VT}(s) = \{Z_T(s) - R_0\}/\{Z_T(s) + R_0\}$

But, $Z_T(s) = 1/sC_L$

$$\therefore \rho_{VT}(s) = \{(1/sC_L) - R_0\}/\{(1/sC_L) + R_0\}$$
$$= -1 + [(2/R_0C_L)/\{s + (1/R_0C_L)\}]$$

A step input starts down the line and is characterized by $V/2s$. However, when it reaches the end of the line at $t = t_d$ it is $U(t - t_d)V/2$, in the time domain.

In Laplace Transform symbolism this is $(Ve^{-st_d})/2s$

The reflected waveform is $\{\rho_{VT}(s)Ve^{-st_d}\}/2s$

However, when this arrives at the beginning of the line, after a further delay t_d, it becomes $\{\rho_{VT}(s)Ve^{-2st_d}\}/2s$.

Substituting for $\rho_{VT}(s)$ gives the expression

$$[-\{Ve^{-2st_d}\}/2s] + Ve^{-2st_d}[1/R_0C_Ls\{s + (1/R_0C_L)\}]$$

Adding the component $V/2s$ gives,

$$v_1(s) = V/2s + [-\{Ve^{-2st_d}\}/2s] + Ve^{-2st_d}[1/R_0C_Ls\{s + (1/R_0C_L)\}]$$

Taking the inverse transform, and remembering that

$$\mathcal{L}^{-1}e^{-bs}F(s) = f(t - b)U(t - b)$$

gives,

$$v_1(t) = [VU(t)/2] + U(t - 2t_d)[V\{1 - \exp - (t - 2t_d)/R_0C_L\} - V/2]$$

This is eqn (5.8) of the text.

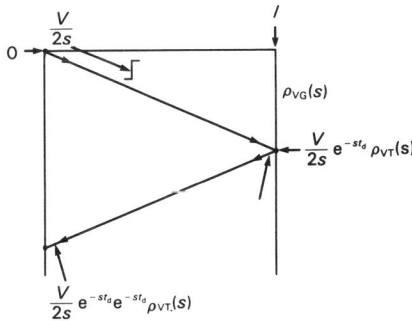

Fig. S5.3

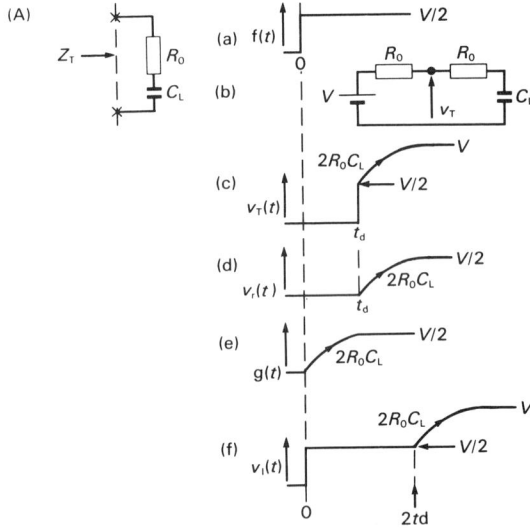

Fig. S5.4A

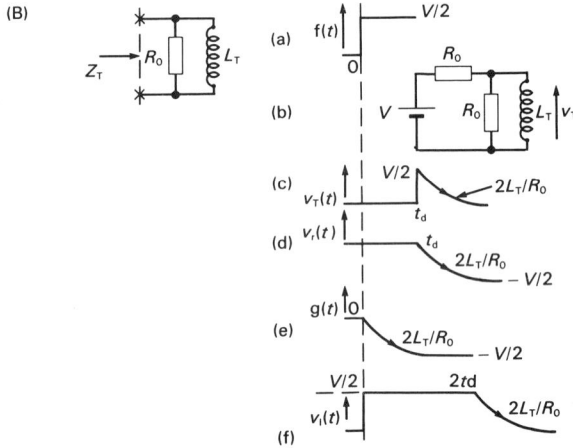

Fig. S5.4B.

P5.4 (A) See Fig. S5.4A
(a) $v_f = f(t)$;   (b) Equivalent circuit for $t = t_d$;
(c) $v_T(t)$;   (d) $v_r(t)$;   (e) $g(t)$;   (f) $v_I(t)$.
(B) See Fig. S5.4B
(a) $v_f = f(t)$;   (b) Equivalent circuit for $t = t_d$;
(c) $v_T(t)$;   (d) $v_r(t)$;   (e) $g(t)$;   (f) $v_I(t)$.
(C) See Fig. S5.4C

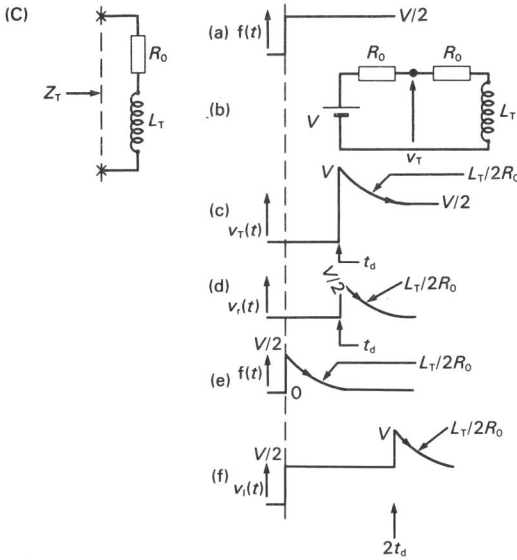

Fig. S5.4C.

(a) $v_f = f(t)$;   (b) Equivalent circuit for $t = t_d$;
(c) $v_T(t)$;   (d) $v_R(t)$;   (e) $g(t)$;   (f) $v_I(t)$.

P5.5   From problem P5.4:

(A)

$$v_I(t) = \{VU(t)/2\} + U(t - 2t_d)V[1 - \exp\{-(t - 2t_d)/2R_0C_L\}]/2$$
$$v_T(t) = U(t - t_d)(V/2)[1 + 1 - \exp\{-(t - t_d)/2R_0C_L\}]$$
$$\text{or, } v_T(t) = U(t - t_d)(V/2)[2 - \exp\{-(t - t_d)/2R_0C_L\}]$$

(B)

$$v_I(t) = \{VU(t)/2\} - \{U(t - 2t_d)V/2\}[1 - \exp\{-(t - 2t_d)R_0/2L_T\}]$$
$$v_T(t) = U(t - t_d)(V/2)\exp\{-(t - t_d)R_0/2L_T\}$$

(C)

$$v_I(t) = \{VU(t)/2\} + \{U(t - 2t_d)V/2\}[\exp\{-(t - 2t_d)2R_0/2L_T\}]$$
$$v_T(t) = \{U(t - t_d)V/2\}[2 - \exp\{-(t - t_d)2R_0/L_T\}]$$

P5.6   Consider the ramp waveform of Fig. S5.6(a), for which $v_P = (Vt/2t_c)$
$= Kt$ (where $K = V/2t_c$), applied to the circuit of Fig. S5.6(b).

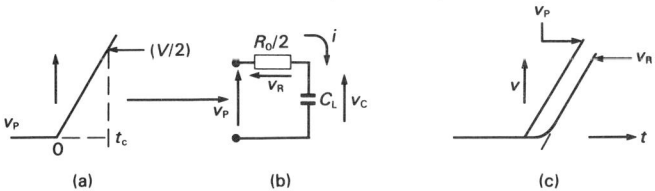

(a)   (b)   (c)

Fig. S5.6.

Now, $i(R_0/2) = v_R = (v_P - v_C)$

and, $i = C_L(dv_C/dt) = C_L d(v_P - v_R)/dt$

Hence, $\left(\dfrac{2v_R}{R_0}\right) = C_L\left(\dfrac{dv_P}{dt}\right) - C_L\left(\dfrac{dv_R}{dt}\right)$

Substituting for (dv_P/dt) gives,

$$\dfrac{2v_R}{R_0} = C_L K - C_L\left(\dfrac{dv_R}{dt}\right)$$

or, $\dfrac{dv_R}{dt} + \dfrac{2v_R}{C_L R_0} = K$

The solution of this is,

$v_R = K(C_L R_0/2) [1 - \exp\{-2t/C_L R_0\}]$: see Fig. S5.6(c).

But, $v_C = v_P - v_R$

$\therefore v_C = Kt - v_R$

or, $v_C = K\{t - (C_L R_0/2)\} + K(C_L R_0/2) [\exp\{-2t/C_L R_0\}]$

If the incident waveform is v_P, the reflected waveform is v_r.

$v_r = v_c - v_P = -KC_L R_0/2 + K(C_L R_0/2) [\exp\{-2t/C_L R_0\}]$

For $t \gg (R_0 C_L/2)$

$|v_r|_{max} = KC_L R_0/2 = (V/4t_c)R_0 C_L$

If the input is not a ramp but a truncated-ramp the same argument holds providing $t_c \gg R_0 C_L/2$. The duration of $|v_r|_{max}$ is of the order of t_c.

CHAPTER 6

P6.1 From eqn (6.9),

$K_{cf} \triangleq \{(C_m Z_0/2) - (L_m/2Z_0)\}$

Since L and C are measured per-unit-length,
$[C_m Z_0] = [C] [\sqrt{L/C}] = [\sqrt{LC}] = [T]$/unit-length
Similarly $[L_m/Z_0] = [\sqrt{LC}] = [T]$/unit-length
From eqn (6.21),

$K_{cr} \triangleq [(C_m Z_0) + (L_m/Z_0)]/4t_u$

We have already shown that the two quantities in the square brackets have the dimensions of time per-unit-length. As t_u is defined as time per-unit-length, K_{cr} is dimensionless.

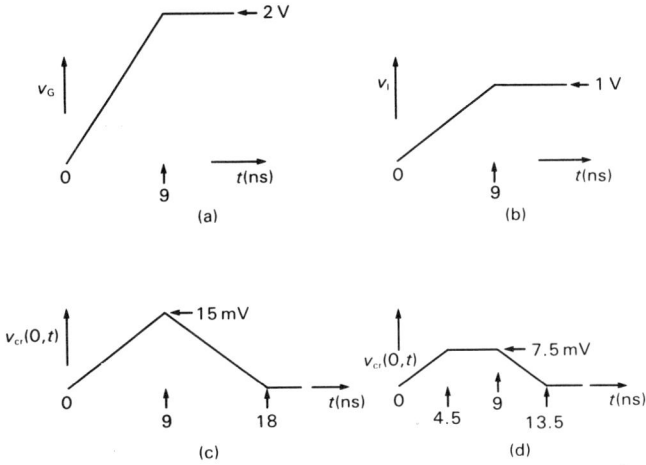

Fig. S6.2.

P6.2 In Fig. S6.2
(a) and (b) are self-explanatory;
(c) gives $v_{cr}(0, t)$ for $t_c = 2t_d$ (compare with Fig. 6.11 of text);
(d) gives $v_{cr}(0, t)$ for $t_c = 4t_d$.

P6.3 $K_{cr} \times 1 \text{ V} = 15 \text{ mV}$: $l/2$ corresponds to $t_d/2$, i.e. 15 ns. See Fig. 6.6, of text, for explanation of waveform duration.

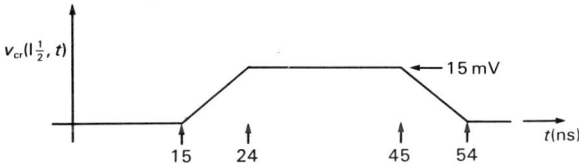

Fig. S6.3.

P6.4 In Fig. S6.4,
   (a) shows $v_{cr}(0, t)$ if line B correctly terminated.
   (b) shows $v_{cf}(l, t)$: this is double the amplitude of the correctly terminated case because of the open circuit ($\rho_{VT} = +1$).
   (c) $v_{cr}(0, t)$ is the sum of (a) and a time-displaced version of (b).

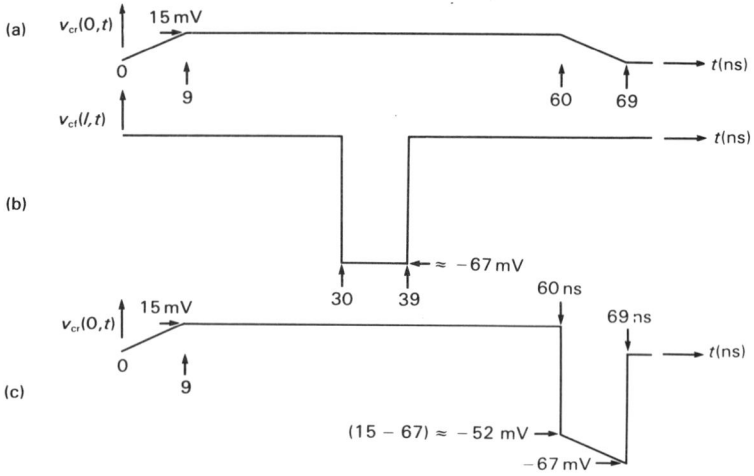

Fig. S6.4.

P6.5

$$v_{cr}(0, t) = K_{cr}[v_1(t) - v_1(t - 2t_d)]$$

Figure S6.5(d) shows $v_{cr}(0, t)$ as comprising constituent parts (a) and (c).

$$v_{cf}(l, t) = K_{cf} \, l \, \frac{d}{dt} \{v_1(t - t_d)\}$$

$$\frac{dv_1}{dt} = + \frac{1}{\tau} \exp(-t/\tau)$$

$$\therefore \; v_{cf}(l, t) = \frac{K_{cf} l}{\tau} [\exp\{-(t - t_d)/\tau\}]$$

At $t = t_d$, $v_{cf}(l, t_d) = -0.18$ V.

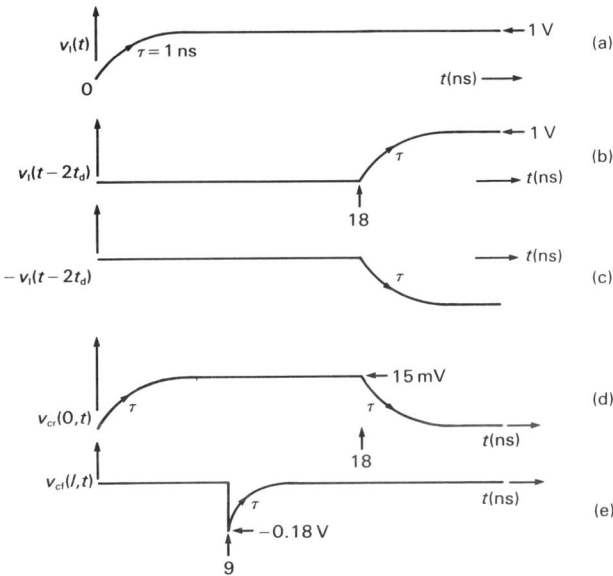

Fig. S6.5.

P6.6 $v_G(t) = V[1 + \tanh\{\alpha(t - t_h)\}]$

This exhibits symmetry about point $t = t_h$ at which $v_G(t) = V$.
Note that for $x \to \infty$, $\tanh x \to 1$ but for $x = 2$, $\tanh x > 0.999$. Similarly, for $x < -2$, $\tanh x < -0.999$. We will assume $\tanh \alpha(t - t_h) = 1$ for $\alpha(t - t_h) = 2$, i.e. $t = t_2 = t_h + (2/\alpha)$ in Fig. 6.6(a). We also assume $[1 + \tanh \alpha(t - t_h)] = 0$ for $\alpha(t - t_h) = -2$, i.e. $t = t_1 = t_h - (2/\alpha)$. As $t_h = 3$ ns and $\alpha = 2$ ns^{-1} (given) the curve for $v_1(t) = v_G(t)/2$ is shown in Fig. S6.6(b).

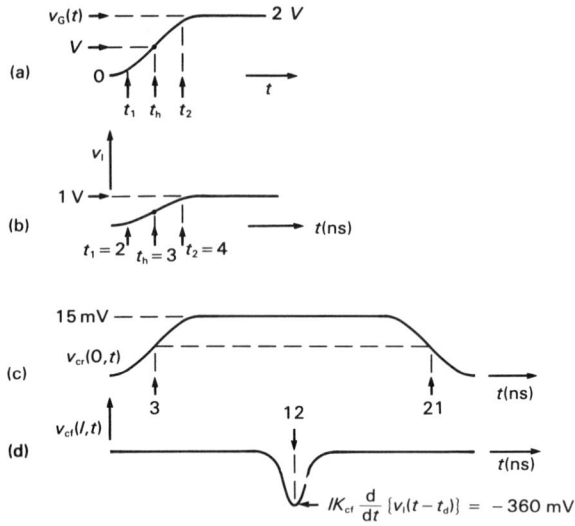

Fig. S6.6

Now, $\dfrac{d}{dt}\{\tanh \alpha(t - t_\mathrm{h})\} = \alpha \operatorname{sech}^2 \alpha(t - t_\mathrm{h})$

$$= \alpha[1 - \tanh^2 \alpha(t - t_\mathrm{h})]$$

The maximum value of $\alpha[1 - \tanh^2 \alpha(t - t_\mathrm{h})]$ occurs at $t = t_\mathrm{h}$ and has the value α. $v_\mathrm{cr}(0, t)$ and $v_\mathrm{cr}(l, t)$ are thus shown in Fig. S6.6(c) and (d) respectively.

P6.7 In Fig. S6.7,
  (a) shows $v_I(t)$; (b) shows $-v_I(t-2t_d)$.
  (c) which is the required $v_{cr}(0, t)$, is the algebraic sum of (a) and (b) multiplied by K_{cr}.
  (d) shows $v_{cf}(l, t)$. This is a differentiated, scaled and time-displaced version of (a).

$$\frac{K_{cf}lV}{2t_c} = -0.01 \text{ V}.$$

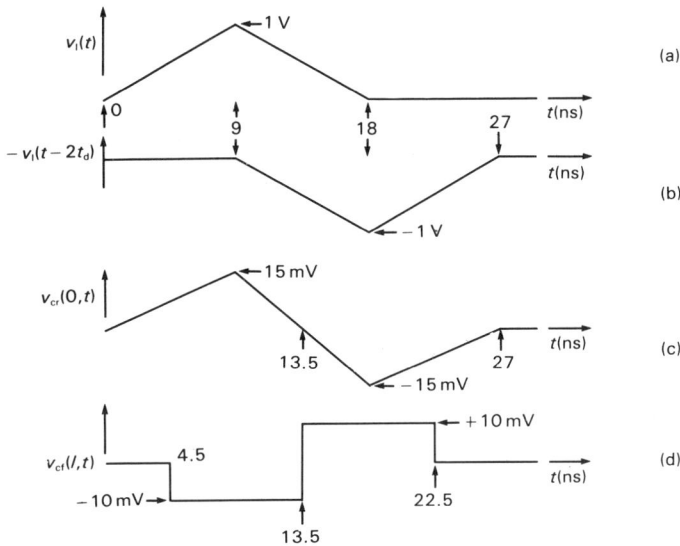

Fig. S6.7.

CHAPTER 7

P7.1 For Cable 2 to be matched, R_3 must be 50 Ω. The impedance to earth at X is R_1 in parallel with $(R_2 + 50\ \Omega)$. Given that $(v_Y/v_X) = 0.5$, it follows that $R_2 = 50\ \Omega$. Hence, $R_1 = 100\ \Omega$.

P7.2 The problem requires that,
$\{R_1 R_2/(R_1 + R_2)\} = 100\ \Omega$ and $\{5R_1/(R_1 + R_2)\} > 3$ V. If we replace '>' by '=', we can substitute into the first equation to obtain $R_2 \approx 168\ \Omega$. Let $R_2 = 150\ \Omega$, a preferred value. Then it follows that $R_2 = 390\ \Omega$, also a preferred value, is suitable. The parallel combination of R_1 and R_2 is then 108 Ω. In practice this would be an acceptable termination for a 100 Ω line.

In the '0' condition the line drive transistor must have a collector–emitter voltage not exceeding 0.4 V. It must thus be able to sink [{(5 – 0.4)/0.15} – {0.4/0.39}] mA in this condition, i.e. approx. 30 mA.

P7.3 Desirable device parameters of Q are: high common-emitter direct current gain, at the maximum operating current I, to reduce loading on the TTL drive gate; a high transition frequency f_T (e.g. >500 MHz) for collector currents in the range 0 to I; a low collector–base capacitance, to reduce capacitive loading on the drive gate.

If the '1' level output of the gate is assumed to be 3.7 V (a typical value) and we allow a base–emitter voltage drop of 0.7 V, nominal, this gives a 3 V signal sent down the 50 Ω line. Thus $I \approx 60$ mA.

The circuit would not be suitable as a high-impedance probe for the detailed observation of fast digital signals because the base–emitter junction of Q does not conduct until $V_{BE} > 0.6$ V. Hence, digital signals having an amplitude less than 0.6 V would not be observed. (Waveforms negative-going with respect to earth would also be ignored.)

There is a problem with this circuit as a line driver. If the terminating resistor were inadvertently shorted, during circuit tests, Q might be destroyed because the collector current would be of the order of the short-circuit output current of the gate multiplied by the transistor current gain. This would cause an excessive power dissipation in Q. This condition can be mitigated by the use of a current-limiting resistor in the collector lead of Q. However, the resistor must be small in value if Q is not to saturate when the gate output produces a '1' on its base terminal.

P7.4 The use of a further double-pole-double-throw switch Sw$_T$ in Fig. S7.4 (compare with Fig. 7.5(a) of the text), allows tri-state operation.

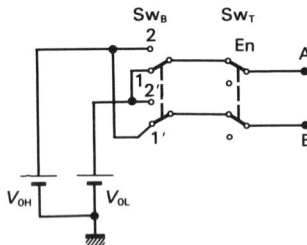

Fig. S7.4.

P7.5 Let $R_G (= 10\ \Omega)$ be the output impedance at x and x'. Matched operation requires that $2(R + R_G) = 100\ \Omega$. Hence, $R = 40\ \Omega$. The practical choice $R = 39\ \Omega$ is appropriate.

P7.6 For balanced drive operation in Fig. P7.6(a),

$$I_{max} = (2.8 - 0.4)\ \text{V}/100\ \Omega = 24\ \text{mA}.$$

For the termination in Fig. P7.6(b),

$$I_{max} = 2.8\ \text{V}/50\ \Omega = 56\ \text{mA}.$$

The termination in (b) would be superior to that of (a) with respect to the common-mode noise-rejection requirements of a receiver connected there. This is because the common-mode impedance to earth at each of the receiver's inputs is less in (b) than in (a), and thus the magnitude of any induced noise spikes would be less.

P7.7 A sketch of a bidirectional repeater is shown in Fig. S7.7. $R_T = R_0$ R_1 and D_1 and R_2 are operative while D_2 are inoperative, and vice versa.

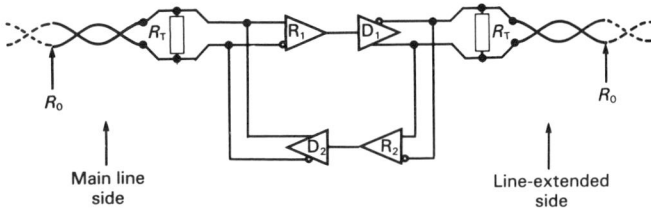

Fig. S7.7.

P7.8 Imagine the line disconnected just to the right of points A, B. Then the conditions which would exist in one state of the current-mode driver switch are shown in Fig. S7.8(a), and the conditions which would exist in the other are depicted in (b). The open-circuit voltage change at A is $(v_{2'} - v_2) = IR$ and this is 'seen' from a source resistance R. Similarly

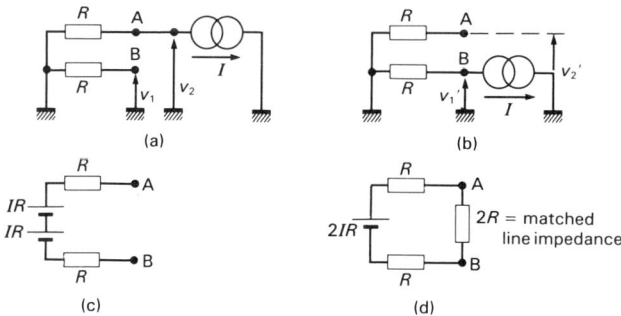

Fig. S7.8.

the open-circuit voltage change at B is $(v_1' - v_1) = -IR$, again 'seen' via a source resistance R. Fig. 7.8(c) shows that the open-circuit voltage change between A and B is $2IR$ seen via a source resistance $2R$. The voltage change propagated down the matched line, when it is connected, can be determined from the equivalent circuit of Fig. S7.8(d). It is IR. (The magnitude of the signal *swing* on each wire is $IR/2$.) But, $R = 50$ Ω; hence for a 1 V change, $I = 20$ mA.

References

The literature on lines and cables is extensive, being scattered through textbooks, learned papers, and handbooks. The following lists make no attempt at being exhaustive; they represent a selection appropriate to the material covered in the text.

BOOKS

1   Johnson, W.C. (1950). *Transmission Lines and Networks*. McGraw-Hill, mainly Chapter 1.
2   Millman, J.M. and Taub, H. (1965). *Pulse, digital and switching waveforms*. McGraw-Hill, Chapter 3.
3   Matick, R.E. (1965). *Transmission lines for digital and communication networks*. McGraw-Hill, Chapter 5.
4   Oliver, B.M. and Cage, J.M. (1971). *Electronic Measurements and Instrumentation*. Inter University Electronic Series 12, 61–4, especially.
5   Coombs, C.F. Jr. (1979). *Printed Circuits Handbook*. McGraw-Hill, New York, 1–29.
6   Harper, C.A. (ed.) (1977). *Handbook of Wiring, Cabling and Interconnecting for Electronics*. McGraw-Hill.
7   Lewin, D. (1985). *Design of Logic Systems*. Van Nostrand Reinhold (UK), 30.

REPORTS/PAPERS
CHAPTER 4

1 Latif, M.A. and Strutt, M.J.O. (1968). Simple graphical method to determine line reflections between high-speed logic circuits. *Electronics Letters* **4** (23), 496–7.
2 Crowther, G.O. (1970). Reflection phenomena when TTL gates are connected to long lines. *Electronic Equipment News.*
3 Hart, B.L. (1972). Graphical analysis of pulses on lines. *Wireless World,* 427–31.
4 Garrett, Lane S. (1970). Integrated-Circuit digital logic families II — TTL devices. *I.E.E.E. Spectrum,* 63–71.
5 *Schottky TTL.* (Undated). Application Report B93. Texas Instruments.
6 *High Speed Comparator Applications.* (1979). Publication PS 1652, The Plessey Co. Ltd, 10–19.
7 *ECL High-Speed Logic* (400 MHz 100 K Range). (1985). Technical Publication 62545003, Mullard Ltd.

CHAPTER 5

1 *Time Domain Reflectometry.* (1964). Application Note AN62, Hewlett-Packard.
2 *Cable Testing with Time Domain Reflectometry.* (1965). Application Note AN67, Hewlett-Packard.
3 Hart, B.L. (April 1971). Demonstrating line pulse reflections. *Electronic Components,* 333–5.

CHAPTER 6

1 Jarvis, D.B. (Oct. 1963). The effects of interconnections on high-speed logic circuits. *IEEE Transactions on Electronic Computers,* 476–87.
2 Feller, A., Kaupp, H.R. and Digiacomo, J.J. (1965). Crosstalk and reflections in high-speed digital systems. *Proc. Fall Joint Computer Conference,* 511–25.
3 DeFalco, J.A. (1970). Reflection and crosstalk in logic circuit interconnections. *IEEE Spectrum,* 44–50.

CHAPTER 7

1   Hart, B.L. (1984). A Schmitt trigger design technique for 'logic-noise-spike' elimination. *IJEEE* **21**, 353–6.
2   *Industry Standard Line Circuits*. (UK 1985). Booklet LL8E. Texas Instruments.

Index

Page references in italics indicate information contained in either a table or a figure.